HENRY D. HIBBARD

MANUFACTURE AND USES OF ALLOY STEELS

Elibron Classics
www.elibron.com

Elibron Classics series.

ISBN 1-4021-5880-7 (paperback)
ISBN 1-4212-7828-6 (hardcover)

This Elibron Classics Replica Edition is an unabridged facsimile of the edition published in 1919 by John Wiley & Sons, Inc., New York.

WILEY ENGINEERING SERIES

The Wiley Engineering Series will embrace books devoted to single subjects. The object of the Series is to place in the hands of the reader all the essential information regarding the particular subject in which he may be interested. Extraneous topics are excluded, and the contents of each book are confined to the field indicated by its title.

It has been considered advisable to make these books manuals of practice, rather than theoretical discussions of the subjects treated. The theory is fully discussed in text-books, hence the engineer who has previously mastered it there, is, as a rule, more interested in the practice. The Wiley Engineering Series therefore will present the most approved practice, with only such theoretical discussion as may be necessary to elucidate such practice.

WILEY ENGINEERING SERIES

MANUFACTURE AND USES OF ALLOY STEELS

BY

HENRY D. HIBBARD

NEW YORK

JOHN WILEY & SONS, Inc.

London: CHAPMAN & HALL, Limited

1919

This monograph is reprinted with the permission of the Bureau of Mines, Department of the Interior, which issued it as a bulletin.

PRESS OF
BRAUNWORTH & CO.
BOOK MANUFACTURERS
BROOKLYN, N. Y.

CONTENTS

CHAPTER IV

MANGANESE STEEL

CHAPTER V

SIMPLE NICKEL STEELS

CHAPTER VI

NICKEL-CHROMIUM STEELS

CHAPTER VII

SILICON STEELS

CHAPTER VIII

HIGH-SPEED TOOL STEELS

CHAPTER IX

CHROMIUM-VANADIUM STEELS

DEFINITIONS

Definitions of terms used throughout this monograph are presented below:

Simple steel, often called "carbon steel," consists chiefly of iron, carbon, and manganese. Other elements are always present, but are not essential to the formation of the steel, and the content of carbon or manganese, or both, may be very small.

Alloy steel is steel that contains one or more elements other than carbon in sufficient proportion to modify or improve substantially and positively some of its useful properties.

Simple alloy steel is alloy steel containing one alloying element, as, for example, simple nickel steel.

Ternary steel is alloy steel that contains one alloying element, the term being synonymous with "simple alloy steel."

Quaternary steel is an alloy steel that contains two alloying elements, such as chromium-vanadium steel.

Complex steel is an alloy steel containing more than two alloying elements, such as high-speed tool steel.

Alloy-treated steel is a simple steel to which one or more alloying elements have been added for curative purposes, but in which the excess of the element or elements is not enough to make it an alloy steel.

Raw steel is steel as cast, either an ingot or casting.

Natural steel is steel in the condition left by a hot-working operation, and cooled in the open air.

Normalized steel is steel that has been given a normalizing heat treatment intended to bring all of a lot of samples under consideration into the same condition.

Annealed steel is steel that has been subjected to an annealing operation.

Hardened steel is steel that has been hardened by quenching from or above the hardening temperature.

Tempered steel is steel that has been hardened and subsequently tempered by a second lower heating.

These definitions are based on the definition of steel that states that steel must be usefully malleable. The definitions of alloy steels do not include effects which are negative, or the prevention or cure of ills which the steel might possess were the alloying element or elements not added.

An iron alloy is not herein considered as useful unless it presents some useful property or modification of a property not offered to the same degree by a simple steel.

The definition of alloy steel given does not agree with that of all writers on the subject of mixtures of iron with other elements than carbon; but it does agree with that of some who have been careful enough when considering the whole range of elements to designate them as alloys, such as silicon-iron alloys and chrome-vanadium alloys, the range covering the useful alloys or steels as well as those in which the alloying element is added for curative purposes and others that have only a scientific interest.

Elements other than carbon may be desired in steel, and therefore be added to or permitted to remain in it for three distinct purposes, as follows:

1. To give the composition desired and to cure in simple steels some ills or defects that the final product might otherwise possess.

2. To make alloy steels. Such elements are manganese, silicon, tungsten, nickel, chromium, vanadium, cobalt, and others of less importance.

3. To make alloys which, though they excite only a scientific interest, form a great part of the whole field of iron metallurgy. Many a one of these alloys would have a commercial value if another alloy were not known that meets particular requirements more satisfactorily either as to efficiency or as to cost or both.

The various additions of the element manganese to iron

illustrate well the three purposes of alloys as above specified. A moderate amount, usually less than $1\frac{1}{2}$ per cent, is added to molten steel made by an oxidation process (pneumatic or open hearth) to prevent red-shortness. A much larger amount is added to make commercial manganese steel, which should contain 11 to 14 per cent of manganese. Outside of these limits, a great number of manganese-iron alloys may be made, most of which have only a scientific value, though the number of useful ones is always liable to be extended as new requirements and methods of manufacture and treatment arise. Simple commercial steels containing between 1.5 and 2 per cent of manganese are made, and manganese steels containing less than 11 per cent manganese are useful for certain purposes; but the manganese content of the great bulk of the steels made is within the limits given.

The total number of possible alloys of iron with varying proportions of other elements is of course practically infinite. So, indeed, is the number of useful alloys, though they form only a small fraction of the whole number.

This monograph deals exclusively with alloy steels, as defined above, in which the alloying element or elements modify directly, positively, and usefully some of the properties of the products.

ACKNOWLEDGMENT

The author desires to make grateful acknowledgment of the cooperation of the following firms or individuals, who are some of the many who have contributed information for this monograph: Pond Machine Tool Co.; C. W. Leavitt & Co.; Bethlehem Steel Co.; Crucible Steel Company of America; Carbon Steel Co.; Firth-Stirling Steel Co.; Carnegie Steel Co.; Peerless Motor Car Co.; Society of Automobile Engineers; Fore River Engine & Ship Building Co.; Tabor Manufacturing Co.; United Steel Co.; Carpenter Steel Co.; Pennsylvania Steel Co.; Cambria Steel Co.; Columbia Tool Steel Co.; Edgar Allen Manganese Steel Co.; Primos Chemical Co.; Henry Disston's Sons Co.; Simonds Manufacturing Co.; Ford Motor Co.; Perfection Spring Co.; Chrome Steel Works; Standard Roller Bearing Co.; Central Railroad of New Jersey; New York Central & Hudson River Railroad Co.; Delaware, Lackawanna & Western Railroad Co., Pennsylvania Railroad Co.; R. R. Abbott; E. O'C. Acker; Westinghouse Electric & Manufacturing Co.; J. A. Mathews; F. V. Sargent; Hans Boker; W. A. Bostwick; B. E. Eldred; Dr. Paul R. Heyl; Illinois Steel Co.; W. M. Stein; and Leonard Waldo.

INTRODUCTION

The object of this monograph is to give briefly information of present value relating to the manufacture and uses of the various commercial alloy steels, with the hope of stimulating the demand for such steels and extending their practical use.

Alloy steels are included in the so-called special steels, but as the latter term is often used in the mills to designate broadly any steels intended for purposes other than those served by the regular product, it has seemed best to use the more specific term of alloy steels in this monograph.

Alloy steels are bringing about a series of revolutions in various industrial fields in which steel plays an important part. Most elements that could be procured in sufficient quantity have been alloyed with iron in various proportions, either alone or in combination with others, in the search for useful alloy steels. Those steels that have gained and maintained for themselves a place in current use are discussed in this monograph. Some of them have had an ephemeral life of usefulness which would no doubt have been prolonged had not some other more satisfactory steel been developed.

Probably the first useful alloy steel was Mushet's self-hardening tungsten tool steel, patented in 1868. Fifteen years later chromium steel, really containing chromium, was struggling for recognition for some purposes, the chief of which was for the manufacture of solid shot for piercing armor. In both of these steels the effect of the alloying element as used was in a way proportional to the amount contained. In 1882 Hadfield made his epoch-making discovery of manganese steel and demonstrated that in iron metallurgy it is not safe to take for granted anything as to the properties of an alloy of iron with other elements, basing one's opinion on past experience and knowledge, and that the effect of an alloying element may not be proportional to its content. The development of useful nickel steels followed in a few years, and the field thus opened has since then been worked by many able and zealous men, with results of great importance and value.

MANUFACTURE AND USES OF ALLOY STEELS

CHAPTER I

LIST OF USEFUL ALLOY STEELS

THE eight alloy steels named below in the chronological order of their introduction are considered to meet this requirement:

1. Simple tungsten steels.
2. Simple chromium steels.
3. Manganese steel.
4. Simple nickel steels.
5. Nickel-chromium steels.
6. Silicon steels.
7. High-speed tool steels.
8. Chromium-vanadium steels.

The first four and the sixth of these are ternary steels, the fifth and eighth are quaternary, and the seventh is of complex composition. Some of these steels may be treated while molten by the addition of a purifying or solidifying element or elements in such a small quantity as not greatly to affect the final properties. Thus a small amount of titanium, aluminum or vanadium may be added to a chromium or nickel steel and hardly appear in the final analysis. Such a result is seen in the alloy-treated steels.

ALLOY-TREATED STEELS

Alloy-treated steels need be only briefly alluded to here. The method of manufacturing all steels made by the oxidation processes involves the presence in suspension or solution of harmful amounts of oxygen or oxides, and before the metal is cast means must be taken to lessen the oxygen or oxide content to or below an allowable maximum. One or more of certain elements having at steel-melting temperatures a stronger affinity for oxygen than iron has, are added to the molten metal. The oxygen leaves the iron to seize such added elements, forming new products insoluble in the iron, which in time are precipitated, gather together, and leave the metal.

Such unfinished steel also contains in solution a quantity of gases which require to be decomposed or kept in solution, for, if not, when the steel is solidifying, part of the gases will leave it and part will be imprisoned in the metal and form gas holes, of the variety commonly called blowholes. The addition of certain elements to the metal tends to prevent the separation of the gases.

Further there is a tendency of certain of the ingredients of steel to collect or segregate in an injurious way in the upper central part of a large ingot or casting, but the addition of certain elements lessens this tendency.

Elements that are added to prevent, minimize, or cure these ills are manganese, which is the most important, silicon, aluminum, titantium, vanadium, and others of less importance. The effects aimed at are therapeutic and though real and valuable, are mostly negative rather than positive; that is, effort is made to cause the steel to be free from some or all of the defects cited.

The proportions of these elements added are, generally speaking, only enough to cure the defects to be removed or counteracted, with a suitable excess to reasonably assure such a result in view of the uncertainties and irregularities of steel making. The excess of any of the elements named has indeed some effect on the final properties of the steel, but not enough to put the product in the class of the alloy steels. Steels so treated are considered as alloy-treated steels, they being simple steels, outside the subject of this monograph, as the alloying elements do not give new or modified properties of important commercial value.

Some alloying elements are added to simple steel in such proportions as to produce only a curative effect, and the product can not definitely be classified as a simple or an alloy steel, as can be done when a rather large excess of the element is added. The elements vanadium and silicon are examples, both being added to cure ills in the steel, and an excess of either causes the physical properties of the steel to vary to some extent. Both are used also in undoubted alloy steels that have unique properties, which would not be anticipated from the observation of the effect of a moderate excess of either in a simple steel.

Crude alloys of the alloying elements that are used as ingredients in steel making and are not useful themselves in their crude state are not herein considered.

THE USES OF ALLOY STEELS

With few exceptions all alloy steels are heat treated for use, the treatment developing in them the high physical properties they are capable of possessing. No general law regarding the effects of heat treatment of alloy steels can be laid down. Some steels when quenched from a high heat are hardened and others are softened, the latter being generally those with the higher contents of certain of the alloying elements. In respect to the effects of heat treatment each steel is considered by itself.

Developments in the manufacture of alloy steel and in the heat treatment of steel have occurred somewhat simultaneously during the past 30 years, and care is needed lest the benefits gained from one be confounded with those afforded by the other. The highest merit is obtained from the adoption of both developments together—that is, the use of heat-treated alloy steels. Usually heat treatment has contributed more to the superior properties of the metal than has the use of alloys.

The alloy steels discussed in this monograph are considered as regards their value for structural, cutting, or electrical purposes.

Steel used for structural purposes is taken to include that used for the stationary as well as the moving parts of structures and machines, including bridges, buildings, vehicles, machine tools (except the cutting tools), armor plate, ships, motors, machinery for winning and working ores involving resistance to abrasion or corrosion, and wire, except electrical wire, and in general all steel not used in the other two fields.

Steel used for cutting purposes includes that employed to form an actual cutting edge and that used in projectiles for war.

Steel for electrical purposes is used in magnets, core steel; non-magnetic articles, and electrical-resistance devices.

No steel suitable in a commercial sense for two of these purposes is made, though some steels might be used for more than one purpose if a better kind for the other specific purpose were not known; thus a fair tool steel might be made of some of the harder structural steels, and a fair magnet might be made of some of the tool steels.

The effects of the alloying elements in alloy steels are various; thus nickel increases the elastic limit as compared to tensility; chromium increases the hardness of quenched steel; and manganese destroys magnetic susceptibility—effects all of which are valuable for certain purposes.

MANUFACTURE OF ALLOY STEEL

Alloy steels are made by any of the steel-making processes, that is, by any of the variations of the pneumatic processes, by the acid or basic open hearth, by the electric furnace, and by the crucible. For each of the various purposes, however, the practice is more limited, the general rule being of course that the cheapest process is employed that will yield a product satisfactory for the purpose in view.

All alloy-steel ingots or castings should be made sound and

with full tendency to pipe. Soundness, which means freedom from gas holes, is, generally speaking, a necessary requirement in order that the product may be sound, as almost any of the alloying elements interferes with or prevents the welding of steel that contains that element so that any contained gas holes in ingots will not be welded up by hot working. Therefore they may, if near the surface, be opened to the air by scaling in heating or by forging and rolling and then be oxidized within and form seams. Chromium and nickel more than other alloying metals prevent welding in steel.

Pipe in the ingot may be shortened in length by casting the ingot with the larger end up, or it may be avoided if the ingot when so cast is squeezed laterally, or if the top is maintained in the molten state until the remainder of the ingot is solid. If neither of these means is employed enough of the top of the ingot must be discarded to get rid of any objectionable pipe. Whether any pipe is permissible depends on the use to be made of the steel, there being many uses for which steel containing a pipe is adapted, or a hole may even be drilled where the pipe would naturally be as is often done to favor heat treatment in massive articles.

The amounts given in this monograph of the different steels produced in the United States have for the most part been determined by indirect means. They are therefore not exact, but are, it is believed, near enough to the truth to warrant presenting them.

The temperatures are given in both centigrade and Fahrenheit degrees.

Alloy steels and other alloys of iron with other elements have been discussed by numerous writers chiefly in their general or purely scientific aspects.

The selected bibliography, at the end of each chapter, relating to the different alloy steels is intended to be limited to articles that bear upon the useful steels. Many of the articles themselves have bibliographies more or less complete. The bibliography is arranged chronologically.

STRUCTURAL ALLOY STEELS

GENERAL CONSIDERATIONS

Structural steels, whose fields of use have already been noted, have some attributes in common, which makes it worth while to consider them collectively to a certain extent. These steels are working great improvements in the production of structures for various purposes, especially where the saving of weight or increase of strength or both are important, the most conspicuous example being undoubtedly the automobile industry. Heat-treated alloy steels with double or treble the strength of the simple steels they replace and with as great or greater reliability are now in regular and most advantageous use. In common with other alloy steels, structural alloy steels owe a part of their superior properties to the presence of the alloying element, but usually far more to heat treatment when it can be given to them. In automobiles the use of alloy steels is generally not advised unless the steels are heat treated, as the gain from their use in the natural or untreated state does not compensate for the increased cost.

Most structural alloy steels are therefore used in the heat-treated condition, when the articles made of them are, like automobile parts, not too bulky or massive. Large pieces like nickel-steel rails and nickel-steel members of bridges are used without heat treatment, the advantages of increased strength and ductility that the metal possesses being due solely to the presence of the alloying element.

The difficulties attending the heat treatment of large steel parts that are bulky for their weight are holding back their general introduction. They require, as nearly as is practicable, to be uniformly heated, uniformly cooled in quenching, and afterward, when cold, to be made true to form, as the quenching operation, however carefully done, usually leaves them warped or twisted. No doubt, in time, means will be found to overcome these drawbacks and such pieces as rails and bridge members of alloy steels will be used regularly in the heat-treated condition. A compact object like an armor plate, though very large, may be quenched without unmanageable warping because of its simple shape. The difficulty in making straight and true such an article as a heat-treated rail of pearlitic alloy steels lies largely in the springiness of the treated metal. It is not easy to give it the correct amount of set needed to counteract or obliterate a crook, bend, or twist that may result from quenching. Yet this is necessary when the piece must be straight or true to shape. Stretching slightly beyond the elastic limit as is done to some thin steel sheets and relatively small bars to straighten them might be efficacious, but is not to be easily done with a piece of such irregular cross-section as a rail.

The effects of heat treatment are so great that a certain steel may be given a very wide range of properties, depending on the treatment, and any desired set of properties within that range may be obtained solely by varying the heat treatment. The principal variant is the degree of the second heating. The lower this is, the stronger and stiffer the steel, and the higher, the weaker and more ductile it is.

This effect of heat treatment on steel is illustrated by a table published by a producer giving the results of 40 tensile tests made from one heat of steel, each test piece having had a different heat treatment. Five, which cover the range, are given in the table below.

RESULTS OF TENSILE-STRENGTH TESTS OF 5 PIECES OF STEEL EACH RECEIVING A DIFFERENT HEAT TREATMENT

Tensility.	Elastic Limit.	Elastic Ratio.	Elongation in 2 Inches.	Contraction of Area.
Pounds.	Pounds.	Per Cent.	Per Cent.	Per Cent.
84,850	50,500	60	28	67.5
120,975	90,000	74.5	14.5	51
166,950	157,500	94	12.5	44
20 ,600	200,000	97	13	48.7
240,975	225,000	93	9	20.5

Analysis of the original steel showed C, 0.25; Mn, 0.50; Cr, 1.07; and V, 0.17 per cent, but similar results could be obtained with a variety of compositions.

For making small parts that must be true and well finished the structural alloy steels are generally heat treated before they are machined, and this requirement prevents the use in such parts of steel of the highest strength attainable, because steel having that strength is not commercially machinable. Generally speaking, any part that is to have an elastic limit of more than 100,000 pounds per square inch must be treated after having been machined, not before, because most steels having a higher elastic limit than that are too hard to allow machining by commercial processes, though chromium-vanadium steels with an elastic limit of 150,000 pounds per square inch are claimed to be machinable, that is, they may be cut with high-speed steels at a profitable rate. An elastic limit of 100,000 pounds or more per square inch can be imparted to steel only by heat treatment, as no untreated steel of a commercial grade will have so high a limit.

Some of the makers of structural alloy steels are publishing for each of their steels a graph showing the physical properties the steel will have when hardened and then drawn to different temperatures. Of course, the graphs give *ex parte* information which is subject to confirmation before acceptance, but the plan is excellent as giving the most information in the least space. Similar graphs of many alloy steels prepared by consumers are expected to be soon available for comparison. From these graphs a new user of these steels may choose the properties he desires and specify the steel he wishes, making some allowance of course (say 10 per cent) for the uncertainties of manufacture and treatment. The steel maker or treater, to be reasonably sure of meeting the requirements, will aim to exceed the properties specified, and the net result will usually be that the steel will have practically the properties desired.

The table of properties given later show how much more the properties possessed are imparted by treatment than by composition.

The size or massiveness of the article has a great effect on the results obtained by any given heat treatment. The greater the mass the lower the qualities, though not in exact proportion. Thus the mass must always be considered in connection with the properties desired, and the composition and heat treatment prescribed must be modified accordingly, though even then the effect of mass may be only partly compensated for.

The modulus of elasticity of many, if not all, structural alloy steels in common with other steels is not changed much by heat treatment or variations in composition* and is usually between 28,000,000 and 30,000,000 pounds per square inch; that is, the modulus of the steel in its annealed, hardened, and tempered condition remains practically unchanged. The following table was compiled from data given by Landau.†

MODULI OF ELASTICITY OF SOME ALLOY STEELS

Composition of Steel.								Modulus.
C	Si	Mn	P	S	Cr	Ni	V	
Per Cent	Per Cent	Per Cent	Per Cent	Per Cent	Per Cent	Per Cent	Per Cent	
0.50	0.13	0.82	0.01	0.02	1.25	.....	0.14	29,240,000
.47	1.83	.70	.01	.01	.....	.....	.....	28,950,000
.48	.16	.44	.01	.01	.98	2.02	.....	28,840,000
.30	.19	.64	.01	.01	.....	3.25	.18	28,260,000
.25	.21	.74	.01	.01	.....	3.55	.....	28,170,000
.24	.21	.46	.01	.02	.96	2.02	.....	28,200,000
.25	.16	.50	.01	.....	1.05	.....	.16	30,158,000

Because of the unchangeability of the modulus of elasticity the stiffness or rigidity of steel within the elastic limit is not changed either by heat treatment or the presence of any of the alloying elements, except perhaps manganese in manganese steel and nickel in high-nickel steels.

Heat treatment does increase the elasticity, however, so that a piece of heat-treated steel may return to its original

* Landau, David, Influences affecting the fundamental deflection of leaf springs: Bull. Soc. Automobile Eng., vol. 5, March, 1914, p. 430.

† Landau, David, op. cit., pp. 431–434.

form after having endured a stress that would have permanently deformed it in its untreated condition; that is, it is given some of the springiness of heat-treated springs.

Many of the structural steels, particularly those used in automobile manufacture, have a great endurance against fatigue when subjected to repeated alternating stresses. The heat treatment increases their durability in this test even more noticeably than it does the properties determined by the tensile test.

SELECTED BIBLIOGRAPHY *

General Articles Dealing with Two or More Alloy Steels

METCALF, WILLIAM. Alloy steels. Proc. Am. Soc. Test. Mat., vol. 4, 1904, pp. 204–214.

GUILLET, LÉON. Les aciers spéciaux. 2 tomes, Paris, 1904–1905; also in Rev. mét., t. l. 1904; t. 2, 1905.

GRENET, L. Note sur la trempe de l'acier. Bull. Soc. l'ind. min., t. 4, sér. 4, 1905, pp. 973–999.

NOBLE, H. Fabrication de l'acier. Paris, 1905, pp. 543–563.

GUILLET, LÉON. Etude industrielle des alliages métalliques. 2 tomes, Paris, 1906.

—— The industrial future of special steels. Iron and Steel Mfg., vol. 11, February, 1906, pp. 89–95.

—— Quaternary steels. Jour. Iron and Steel Inst., 1906, pt. 2, pp. 1–141.

HOWE, H. M. Iron, steel, and other alloys. 2d ed., 1906.

LEDEBUR, A. Handbuch der Eisenhüttenkunde. Leipzig, 1906, pp. 348–399.

CAMPBELL, H. H. The manufacture and properties of iron and steel. 4th ed., 1907, pp. 343–391. Describes the influence of certain elements on the physical properties of steel.

FAY, T. J. Special auto steel. Proc. Am. Soc. Mech. Eng., vol. 28, June, 1907, pp. 1745–1777.

HAYNES, ELWOOD. Material for automobiles. Trans. Am. Soc. Mech. Eng., vol. 29, 1907, pp. 209–216.

LAKE, E. F. Heat treatment of alloyed steels. Am. Machinist, 1907, vol. 30, pp. 152, 196, 289.

—— Steels used in automobile construction. Am. Machinist, vol. 30, March 14, 1907, pp. 376–382.

* This bibliography and those at the end of subsequent chapters were compiled by the United Engineering Library, 29 West Thirty-ninth Street, New York City. The references are arranged chronologically.

Guillet, Léon. Special alloy steels and their mechanical applications. Eng. Mag., vol. 36, October, 1908, pp. 65–75.

Revillon, Louis. Les aciers spéciaux au salon de l'automobile. Rev. mét., t. 5, February, 1908, pp. 53–68.

Turner, W. L. The static and dynamic properties of steels. Iron Age, vol. 82, July 2, 1908, pp. 53–55.

Giesen, Walter. The special steels in theory and practice. Carnegie Scholarship Mem., Iron and Steel Inst., vol. 1, 1909, pp. 1–59.

Guillet, Léon. Les aciers spéciaux industriels. Rev. mét., t. 6, June, 1909, pp. 810–813.

Law, E. F. Alloys and their industrial applications. London, 1909, 269 pp.

Longmuir, Percy. High-tension steels. Jour. Iron and Steel Inst., pt. 1, 1909, pp. 383–403. Gives results of tests of nickel steels, nickel chromium, and chromium vanadium steels.

Mathews, J. A. Alloy steels for motor-car construction. Jour. Franklin Inst., vol. 167, May, 1909, pp. 379–397.

—— The nature and treatment of alloy steels. Iron Age, vol. 83, Jan. 7, 1909, pp. 24–26.

Portevin, A. M. Contribution à l'Etude des aciers spéciaux ternaires. Rev. mét., t. 6, December, 1909, pp. 1164–1303. Carnegie Scholarship Mem., Iron and Steel Inst., vol. 1, 1909, pp. 230–264. Includes bibliography.

Guillet, Léon. Special steels. Iron Trade Rev., vol. 46, Feb. 3, 1910, pp. 367–369. Translated from a paper read before the International Association for Testing Materials.

Haenig, A. Der Konstruktionstahl und seine Mikrostruktur, unter besonderer Berücksichtigung des modernen Automobilstahls. Berlin, 1910.

Souther, Henry. The selection and treatment of alloy steels for automobiles. Jour. Franklin Inst., vol. 170, December, 1910, pp. 437–450.

—— Specifications for materials—specifications for steel. Trans. Soc. Automobile Eng., vol. 5, 1910, pp. 168–175.

—— Notes and instructions referring to materials specified hereinbefore. Trans. Soc. Automobile Eng., vol. 5, 1910, pp. 176–202.

—— Alloy steels for railroad use. Am. Eng. and R. R. Jour., vol. 85, March, 1911, pp. 95–98. Abstract of paper read at joint meeting of the Engineering Club and Altoona Railroad Club.

Harbord, F. W., and Hall, J. W. The metallurgy of steel. 4th ed., London, vol. 2, 1911, pp. 387–417. Discusses special steels or steel alloys.

Lake, E. F. Alloy steel for automobiles. Automobile, vol. 24, June 29, 1911, pp. 1436–1440.

Sandelin, Folke. Specialstal för Konstruktionsändamål. Stockholm,

1911, 173 pp. Includes a bibliography. Paper printed also in Jern Kontorets Annaler, vol. 66, 1911, p. 524.

Stoughton, Bradley. The metallurgy of iron and steel. 1911, pp. 389–411, 514–516. Discusses alloy steels and presents a bibliography.

International Railway Congress. The use of steel; special steels. Bulletin, vol. 25, August, 1911, pp. 849–881. Report and discussion.

Edwards, E. T. Recognizing alloy steels by ingot structure. Iron Age, vol. 89, Feb. 15, 1912, pp. 399–402.

Transactions of the Faraday Society. Magnetic properties of alloys; symposium. Vol. 8, 1912, pp. 94–219.

Mars, G. Spezialstähle; ihre Geschichte, Eigenschaften, Behandlung und Herstellung. Stuttgart, 1912.

Robin, Felix. Traité de métallographie. Paris, 1912.

Sauveur, Albert. The metallography of iron and steel. 1912. 386 pp.

Society of Automobile Engineers. Fourth report of the iron and steel division. Revised notes and instructions referring to materials specified. Transactions, vol. 8, pp. 2, 1913, pp. 7–50. Includes a bibliography.

Hudson, O. F., and Bengough, G. D. Iron and steel; an introductory textbook for engineers and metallurgists. London, 1913.

Del Mar, Algernon. Wrought iron and steel for stamp-mill parts. Min. and Eng. World, vol. 40, April 11, 1914, pp. 687–693.

Hall, J. H. The steel foundry. 1914. 271 pp.

Landau, David. Influences affecting the fundamental deflection of leaf springs. Bull. Soc. Automobile Eng., vol. 5, March, 1914, pp. 428–464.

CHAPTER II

SIMPLE TUNGSTEN STEEL

MUSHET'S air-hardening steel, the first of the alloy steels, may be considered as a simple tungsten steel, though it contained so much manganese (about 2 per cent) that it might with some reason be classed as a quaternary steel, as it contained also about 2 per cent of carbon and 6 per cent of tungsten. The manganese was essential to give the self-hardening property. The steel is now becoming obsolete, having had a useful career and having formed an indispensable link in the development of the high-speed tool steels discussed later.

Tungsten is very heavy, its specific gravity being, according to recent determinations, about 19.3, and it is the most infusible substance known except carbon and, perhaps, boron. These properties have some effect on the production of tungsten steel.

MANUFACTURE OF TUNGSTEN STEEL

Tungsten steel is generally, if not always, made by the crucible process. The pots are charged cold by packing in the materials, the tungsten being placed at the top to counteract in a measure its tendency to settle because of its high specific gravity. If this tendency operated unchecked there might be at the bottom of the pot a rather infusible mush of high-tungsten alloy which would not pour out, and if it did the ingot would have an irregular composition because of the uneven distribution of the tungsten.

The steel is melted and then "killed" in the crucibles by holding them in the furnace for 30 or 40 minutes after the charge has melted, until the steel ceases to bubble or work and lies dead in the pot.

The pots are sometimes cast singly or doubly by hand pouring or collectively by means of a ladle into which all the pots of a furnace charge are emptied. Good tungsten steel makes remarkably sound solid ingots, except for the pipe, though tungsten itself is not considered to aid in removing or controlling either the oxides or the gases. It is added solely for its effect on the finished and treated steel.

This lack of power of tungsten to deal with oxides and gases arises no doubt from its low calorific power, its heat of combustion being given (with qualification) as about 1,000 calories, whereas iron burned to Fe_3O_4 gives 1,612 calories.

METHOD OF WORKING

Simple tungsten steels of commercial grades are heated, forged, and rolled in much the same manner as other high-carbon steels, presenting no special problems or difficulties.

PROPERTIES AND USES

Simple tungsten steel is at present chiefly used in permanent magnets for electric meters, in small dynamos, and hand use, for which it has been used for 30 or 40 years. The consumption in 1913 is thought to have been between 5,000 and 6,000 tons. This steel contains about 0.6 per cent of carbon and 6 per cent of tungsten. Some has been made in recent years containing 0.2 to 0.3 per cent of vanadium, chromium, or molybdenum which were considered at the time to give greater retentivity to the steel, but those ingredients are now generally held to be of no practical value, adding nothing to the fitness of the steel for its purpose.

Some buyers of magnet steel do not specify composition but only performance, that is, what magnetic properties the steel must have.

To make permanent magnets retain their magnetism as much as possible they are made very hard by heating and quenching. They are then magnetized, and if they are to be used for electric meters they are seasoned by a treatment involving protracted heating to 100° C. (212° F.) to make their magnetism as nearly constant as possible.

A variety of tungsten steel containing about 1 per cent of carbon and 3 to 4 per cent of tungsten is made and used as a tool steel for taking finishing cuts on iron and steel in the machine shop. It acts more like a simple steel than a self-hardening steel, as it requires to be hardened by quenching in water and then drawn in the same general way that simple steels have been drawn, presumably for thousands of years. It will cut at a higher speed than a simple steel, say 40 feet per minute on steel having a tensile strength of 80,000 pounds per square inch, and is also more durable.

The presence of tungsten in steel is generally stated to lower the fusion point of the steel. Mars* gives a table of fusion points of tungsten steels with contents of tungsten ranging from 0.5 to 17 per cent, from which he concludes that tungsten lowers the fusion point. However, when his results are corrected for the lowering effects of the contained carbon, silicon, and manganese doubt arises as to the correctness of his conclusion. Thus, a steel containing 0.66 per cent C, 0.03 per cent Si, 0.04 per cent Mn, and 3.11 per cent W fused at 1,488° C. The carbon would lower the fusion point about 60° C., and the silicon and manganese slightly, so that the plain iron-tungsten alloy should have a fusion point a little above 1,548° C., which is about 20° C. above that of pure iron. Seemingly this is the effect of 3.11 per cent tungsten.

The erosion of the bore of cannon by the powder gases is held to depend largely on the fusion point of the metal of the tube or liner, the higher the point the greater being the resistance to erosion. So it has been found that the nearer the metal comes to being pure iron the higher its fusion temperature and the better it resists erosion, but the strength required compels a certain amount of hardening and strengthening elements to be present in the steel. Tungsten raises the strength and pos-

* Mars, G., Die Specialstähle, ihre Geschichte, Eigenschaften, Behandlung und Herstellung. Stuttgart, 1912.

sibly the temperature of fusion and so has been employed for the tubes of cannon, particularly by the Government of Austria. Arnold and Read* found that steel with 0.71 per cent carbon and 5.4 per cent tungsten had in the annealed state a tensility of 88,900 pounds per square inch, an elastic limit of 60,200 pounds, an elongation of 20 per cent, and a contraction of area of 34.7 per cent, values that compare favorably with those of the steels usually employed in the manufacture of cannon.

They give data regarding a series of annealed tungsten steels as follows:

The strength and hardness of these steels may be greatly increased by heat treatment involving quenching and with only relatively small decrease in ductility.

THEORY OF TUNGSTEN STEEL

Arnold and Read concluded that the carbon in the steels they examined was combined with iron when the tungsten was low, but that the higher the tungsten the more of the carbon was combined with it until in steel containing 11.5 per cent of tungsten none of the carbon was combined with iron, but all of it with tungsten. With still higher tungsten content the excess of tungsten was combined with iron.

BIBLIOGRAPHY ON TUNGSTEN STEEL

FORBES, DAVID. Quarterly report on the progress of the iron and steel industries in foreign countries. Jour. Iron and Steel Inst., vol. 2, 1872, pp. 255–294. Gives chemical analysis of Mushet's "special steel" or tungsten steel.

HADFIELD, R. A. Alloys of iron and tungsten. Jour. Iron and Steel Inst., 1903, pt. 2, pp. 14–118. Presents bibliography of tungsten and tungsten steels dating from 1590.

GUILLET, LÉON. Aciers au tungstène. Rev. mét., vol. 1, 1904, pp. 263–283.

—— Constitution et propriétés des aciers au tungstène. Comp. rend., t. 139, Sept. 26, 1904, pp. 519–521.

* Arnold, J. O., and Read, A. A., The chemical and mechanical relations of iron, tungsten and carbon, and of iron, nickel and carbon: Proc. Inst. Mech. Eng., pt. 2, March–May, 1914, pp. 223–248.

DATA REGARDING ANNEALED TUNGSTEN STEELS

Composition.							Tensile Properties.				Condition when Turned.
C	W	Si	Mn	P	S	Al	Tensility.	Yield Point.	Elongation in 2 Inches.	Contraction of Area.	
Per Cent	Per Cent	Per Cent	Per Cent	Per Cent	Per Cent	Per Cent	Pounds.	Pounds.	Per Cent	Per Cent	
0.73	2.4	0.11					84,200	48,100	20.5	31.5	Moderately tough
.71	5.4	.11					88,900	60,200	20.0	34.7	Tough [5]
.70	9.7	.04					126,100	90,000	14.0	22.1	Very tough
.73	15.0	.03	[1]	[2]	[3]	[4]	98,500	....	25.0	43.3	Very tough
.72	21.1	.06					104,300	57,300	20.5	39.2	Very tough, slightly hard
.67	26.3	.06					110,600	....	9.0	11.4	Very tough, slightly hard

[1] 0.15 or less. [2] 0.02 or less. [3] 0.04 or less. [4] 0.01 or less. [5] Tough means that the lathe chips curled off in spirals.

—— Recherches sur les aciers au tungstène. Génie civil, t. 45, 1904, pp. 7, 27.

SWINDEN, THOMAS. Carbon-tungsten steels. Jour. Iron and Steel Inst., pt. 1, 1907, pp. 291–327.

—— The constitution of carbon-tungsten steels. Jour. Iron and Steel Inst., 1909, pt. 2, pp. 223–256.

MARS, G. Magnetstahl und permanenter Magnetismus. Stahl und Eisen, Bd. 29, 1909, pp. 1673, 1769.

ARNOLD, J. O., and READ, A. A. The chemical and mechanical relations of iron, tungsten, and carbon, and of iron, nickel and carbon. Proc., Inst. Mech. Eng., March–May, 1914, pp. 223–279.

JARDINE, ROBERT. Valves of tungsten steel. Autocar, vol. 33, July 18, 1914, pp. 127–129.

CHAPTER III

SIMPLE CHROMIUM STEEL

Simple chromium steels, though one of the earliest if not the first of the alloy steels to be made, are not now largely used, though they are for quite a variety of purposes. In combination with other alloying elements, however, chromium is still one of the most important constituents of alloy steels.

The production of simple chromium steel ingots is thought to have been about 6,000 tons in 1913 with a content of chromium of 0.4 to 2 per cent. It is made by either the acid open hearth or crucible process. If made in crucibles the chromium in the form of ferrochromium is made a part of the original charge, but if made in the open-hearth furnace the ferrochromium is added just long enough before casting for the alloy to be melted and well mixed with the charge.

The consumption of ferrochromium for steel making in this country in 1913 was between 3,000 and 4,000 tons, containing 60 to 70 per cent of chromium. If 1 per cent be assumed as the average content of chromium in the structural steels, and 4 per cent in the high-speed tool steels, a total of about 204,000 tons of steel of all kinds in the ladle contained chromium. The bulk of such steel was quaternary or complex steel, about 6,000 tons being considered as simple chromium steel, as already stated.

At steel-melting temperatures chromium has a greater affinity for oxygen than has iron, and therefore if any oxygen either free or in the form of oxide of iron comes in contact with melted steel that contains chromium, some of the chromium is wasted. It is not oxidized as rapidly as silicon and manganese, however, and therefore has little deoxidizing effect when added to an oxidized iron bath, the oxides and gases being controlled by other means.

The effect of a chromium content up to a maximum of $2\frac{1}{2}$ per cent in steel is to increase the hardness moderately when the steel is in the natural state, as defined on page ix, and particularly when it is in the hardened condition after having been quenched.

METHODS OF WORKING

Chromium steels are cast, forged, and rolled by the same plant and by the same methods as simple steels of the same or slightly higher carbon contents. Castings are annealed, or heat treated, as the conditions warrant or require to give the most suitable properties for the proposed use.

Chromium steels are perhaps never used in the untreated condition, and their properties in that state are therefore not given.

COMPOSITION AND PROPERTIES OF HEAT-TREATED SIMPLE CHROMIUM STEELS

Sample No.	Constituents. C	Mn	Si	S	P	Cr	Tensile Strength	Elastic Limit.	Contraction of Area.	Elongation in 2 Ins.	Ball Hardness.	Heat Treatment. Temperature at which Steel was Quenched in Water	Temperature at which Hardness was Drawn in Air.
	%	%	%	%	%	%	Lbs.	Lbs.	%	%		°C.	°C.
1	0 70	0.54	0.09	0.01	0.01	0.70	129,000	121,700	60	21	235	816	593
2	.70	.54	.09	.01	.01	.70	110,900	105,300	63	26	195	816	649
3	.70	.54	.09	.01	.01	.70	88,000	73,000	68	36	168	816	754
4	.40	.78	.54	.02	.01	.92	143,500	131,600	56	18	242	816	538
5	.40	.78	.54	.02	.01	.92	103,200	90,200	69	26	201	816	714
6	.91	.35	.08	.03	.01	.91	96,800	69,300	63	28	175	....	....

USES OF SIMPLE CHROMIUM STEELS

The longest established use of chromium steels now current is in stamp shoes and dies for pulverizing certain gold and silver ores. These shoes and dies contain 0.8 to 0.9 per cent of carbon, with 0.4 to 0.5 per cent of chromium. They are preferably annealed to destroy ingotism and so impart some toughness to the metal, which increases their durability in an important degree.

Another long-established use of chromium steel is in 5-ply plates for the manufacture of safes for the safe-keeping of valuables. These plates are made of five alternate layers, two of chromium steel and three of soft steel or wrought iron, and after having been hardened offer great resistance to the drilling tools employed by burglars. The plates are, however, necessarily rather thin, usually between one-half and 1 inch thick, so that a safe wall or door to be more than an inch thick must be formed of two or more thicknesses, which are fastened together with screws. Many portable safes so made have been robbed by burglars who used liquid nitroglycerine, which was flowed into the joints and exploded, blowing out the door or wall. Such safes are therefore not made in such great numbers as formerly, but considerable quantities of chromium steel are used in large stationary safes, usually called vaults, where the individual pieces are larger and the other safeguards against burglary so effective that they are not attacked by burglars.

Hardened chromium-steel rolls having 0.9 per cent of carbon and 2 per cent of chromium are used for cold-rolling metals. They are glass hard so that the ball hardness can not be determined, the ball making no impression. The hardness, as determined by the sclerescope, is 107.

Files of chromium steel are excellent, and in 1913 about 3,000 tons of this steel went into them, the carbon content being 1.3 to 1.5 per cent and the chromium content about 0.5 per cent.

An important use of chromium steel is in balls and rollers for bearings. One large maker uses steel containing carbon, 1.10 per cent; chromium, 1.40 per cent; manganese, 0.35 per cent; sulphur, 0.025 per cent; and phosphorus, 0.025 per cent. Sizes smaller than one-half inch diameter are heat-treated by being quenched in water from 774° C. (1,425° F.) and then drawn to 190° C. (375° F.) for half an hour. For larger balls the quenching temperature is 802° C. (1,475° F.). The second heating does not produce an oxide color, but is enough to let down in some degree the internal stresses due to the irregular cooling of quenching so that the balls are less liable to crack spontaneously or to be broken in use.

The strength of a good, well-treated ball is prodigious, a ball three-fourths of an inch diameter, tested by the three-ball method, sustaining a load of 52,000 pounds. On the small area of contact the intensity of the pressure amounts to over one million pounds per square inch.

The Society of Automobile Engineers recommends less chromium than that given above, or 1 to 1.2 per cent. A steel maker recommends 1.5 per cent.

A former important use of these steels was in armor-piercing projectiles. Cubillo* gives as the analysis of such a steel: C 0.85 per cent, Mn 0.38 per cent, Cr 2.31 per cent, Si 0.16 per cent, and P 0.02 per cent. The physical properties were as follows:

PHYSICAL PROPERTIES OF CERTAIN ARMOR-PIERCING STEELS

Treatment.	Elastic Limit.	Tensility.	Elongation in 2 Inches.
	Pounds.	Pounds.	Per Cent.
Annealed	47,700	102,000	20.5
Heat-treated	79,400	114,500	18.5

The greater part of projectiles, however, contain some nickel also, as noted in the discussion of nickel-chromium steels.

Chromium steels are preferred by some builders in certain automobile parts, though most makers of cars prefer nickel chromium.

* Cubillo, L., The manufacture of armor-piercing projectiles: Jour. Iron and Steel Inst., vol. 88, 1913, p. 251.

BIBLIOGRAPHY ON CHROMIUM STEEL

BRUSTLEIN, H. A. On chrome, pig iron, and steel. Jour. Iron and Steel Inst., 1886, pt. 2, pp. 770–778. Briefly describes experiments made by French railway companies.

HADFIELD, R. A. Alloys of iron and chromium, including a report by F. Osmond. Jour. Iron and Steel Inst., 1892, pt. 2, pp. 49–175. Includes discussion and bibliography.

GUILLET, LÉON. Les aciers au chrome. Rev. mét., t. 1, 1904, pp. 155–183.

—— Recherches sur les aciers au chrome. Génie civil, t. 44, 1904, pp. 281–298.

BRUSTLEIN, A. Comment l'acier chromé est devenu un produit industriel. Rev. mét., t. 2, 1905, pp. 508–515.

PERAR, U. C. Steel for cutters, punches, and dies. Am. Machinist, vol. 33, pt. 1, Jan. 6, 1910, pp. 17–20.

ARNOLD, J. O., and READ, A. A. The chemical and mechanical relations of iron, chromium, and carbon. Jour. Iron and Steel Inst., 1911, pt. 1, pp. 249–268.

PORTEVIN, A. M. Sur les aciers au chrome. Rev. mét., vol. 8, 1911, pp. 802–803.

FRIEND, J. N. The corrodibility of nickel, chromium, and nickel chromium steels. Jour. Iron and Steel Inst., pt. 1, 1913, pp. 388–398. See also general bibliography.

CHAPTER IV

MANGANESE STEEL

MANGANESE steel in the commercial meaning of the name is a variety of iron containing 11 to 14 per cent of manganese and 1.0 to 1.3 per cent of carbon. The original patents covered alloys of iron with 7 to 30 per cent of manganese but the steels within the limits of composition given have the greatest strength and ductility of any and are always meant when manganese steel is ordered without further qualification. Departure from these limits of analysis means in a general way that the steel is to that extent unsuited for the structural purposes for which manganese steel has been found applicable.

The bulk of the manganese steel made at present is put into castings, of which about 36,000 tons was made in 1913. The use of hot-worked (rolled or forged) manganese steel is now of importance, some 3,000 tons having been made in 1913, nearly all of which went into rails. These quantities call for about 60,000 tons of molten steel in the ladle.

Though about 40 concerns offer manganese-steel castings, only about half a dozen make any considerable quantity, having supplied practically all of that made in 1913.

MANUFACTURE

Manganese steel is still made in the ladle according to Hadfield's expired patents by the mixture of decarburized iron and 80 per cent ferromanganese. The decarburized iron is prepared either by the pneumatic process, being blown in some one of the many modified pneumatic converters, or in the Siemens furnace. As ferromanganese forms such a large proportion of the charge, about one-seventh, it must be melted or nearly so before being added to or mixed with the decarburized iron, or the resulting steel would be too cold. The ferromanganese is usually melted in graphite crucibles or pots which is an expensive way but not yet generally superseded. The pots are heated in ordinary coal-fired melting holes. They usually may be used to melt eight or ten charges, whereas in steel melting they average less. This longer life when melting ferromanganese comes from both the lower melting temperature needed and the absence of solution of the graphite of the pot by the metal, common ferromanganese being saturated with carbon, whereas molten steel eagerly dissolves any carbon with which it comes in contact. The ferromanganese need not be completely melted, which of course would not be permissible with the steel.

The decarburized iron used should have a low carbon content, not over 0.10 per cent, so that the resulting steel may not have more than is desired. The proportion of carbon to manganese in the steel will, therefore, be a little more than in the ferromanganese.

After the manganese steel has been made in the ladle it should be cast as soon as practicable if it is to be used for castings, but if it is to be used for ingots a little time should be allowed for the silicate formed within the metal to collect and float to the top.

The quantity of manganese is proportioned to the size of the charge of decarburized iron with allowance for loss through oxidation of an amount equal to about $1\frac{1}{2}$ per cent of the steel. Thus 14 per cent is added to yield 12.5 per cent in the steel. Everything is weighed—the stock before melting, the iron added to the vessel if decarburized by the pneumatic process, the molten ferromanganese, which is put in the ladle first, and the decarburized iron are all weighed, the weight of the decarburized iron giving the weight of the molten steel in the ladle. When the decarburized iron is prepared in a tilting Siemens furnace, a part may be taken out at a time and made into steel, the amount of iron being weighed by a suspended weighing machine on the crane.

Earnest efforts have been made to melt ferromanganese in a cupola because of the high cost of melting in pots, and some has been so melted in a commercial way. When the ferromanganese is melted in a cupola there is more loss of manganese, which offsets, in part at least, the lower cost of melting. The loss is diminished to a large extent by charging the ferromanganese in the center of the cupola with coke around the outside so that the metal is largely protected from free oxygen of the blast, which would oxidize manganese if it came in contact with it. It is a matter for calculation which is to be chosen for any given conditions as being the cheaper or, all things considered, the most advantageous.

Another effect of the waste of manganese from the ferromanganese used in making manganese steel is that the carbon content is higher, and perhaps undesirably so in the finished steel, because of its concentration due to the waste of manganese. Carbon is not oxidized under usual conditions in fused ferromanganese, because of the presence of the large excess of manganese, which seizes any oxygen that reaches the alloy.

Manganese steel is usually made and handled in clay-lined ladles. Basic-lined ladles present some advantages, but clay is used because it is so much cheaper and more easily made into a lining. Such a lining is of course to be classed as acid, and sometimes strongly so, when quartz in the form of sand or ganister is mixed with the clay. Molten manganese steel in an acid-lined ladle continually makes a fusible and liquid slag, which is chiefly a silicate of manganese, green in color. It is formed of oxide of manganese, which is continually being oxidized by atmospheric oxygen, and the ladle lining. A little green silicate rises and joins the slag from within the metal, being formed from the silica contained in the decarburized iron and the oxide of manganese formed by substitution of manganese for iron in the oxide of iron in the metal. The thin fluid slag is troublesome, as it is not easily kept from entering the molds with the steel, where it is likely to cause a defect in the ingot or casting.

To avoid the trouble from the very liquid slag a "coffee-pot" ladle is sometimes used, which is provided with a spout like that of a coffeepot, attached at one side and connecting with the ladle interior at the bottom. This arrangement acts effectively to skim the slag, which can not enter the spout until nearly all the steel has been poured out of the ladle.

The corrosion of the ladle lining is deepest at the slag line when the ladle is full, as the slag has more time to act on the lining at that level than at any other.

Because of its large content of carbon, silicon, and manganese, the latter fusing at 1,260° C., manganese steel melts at about 1,325° C., a temperature lower than that of simple steel, and one that favors the running of intricate castings. For the same reason manganese steel, containing so much gas solvent, is usually free from gas holes; but if the decarburized iron of which it is made is too hot, and therefore too heavily charged with gases, the solvent powers of the silicon and manganese may be exceeded, and the steel be saturated with gases, the ingots or castings being consequently infested with blow-holes by the gases liberated in cooling.

COMPOSITION

In making manganese steel one composition is practically standard. The usual analyses of manganese steel lie between the following limits: Carbon, 1.0 to 1.3 per cent; silicon, 0.3 to 0.8 per cent; manganese, 11.0 to 14.0 per cent; phosphorus, 0.05 to 0. 08 per cent. The sulphur content is so low as to be negligible in manganese steel as in other iron-manganese alloys, from which any sulphur that may get in is quickly eliminated by the manganese, probably as sulphide, which rises to the surface or enters the slag. In the open the sulphur is burned by the oxygen of the air, forming sulphurous acid (SO_2), which may often be smelled coming from the molten steel.

Low-manganese steels with 7 to 8 per cent of manganese are finding some use, having a higher and better defined elastic limit than the regular grade and yet with considerable though much less ductility. They are also cheaper to make. They do not flow under impact as freely as the regular commercial grades, because, no doubt, of their martensitic structure.

Manganese-iron alloys containing 3 to 10 per cent of manganese and 1 per cent of carbon are martensitic. With the manganese over 10 per cent the structure is austenitic. The steels having 7 to 10 per cent of manganese are so different from commercial manganese steel that another name should be given them to avoid confusion. The name "loman steel," an abbreviation of "low-manganese steel," has been applied to them and seems to be suitable as a short distinctive name.

GENERAL PROPERTIES

Manganese steel is a hard self-hardening steel, owing this property to its composition and not to treatment. It can not be softened by heating followed by slow cooling. It is, for a metal, a poor conductor of electricity.

Manganese steel has a high coefficient of expansion, small patterns being made with a shrinkage of five-sixteenths of an inch to 1 foot, which sometimes is not quite enough. A shrinkage of five-sixteenths of an inch to 1 foot gives a mean coefficient of expansion of about 0.000024 per degree centigrade.

In respect to specific gravity manganese steel is not to be distinguished from simple steels of the same carbon content, as all have, generally speaking, about the same.

Perhaps the most remarkable property of manganese steel is its almost total lack of magnetic permeability and susceptibility. This metal, containing 85 per cent of iron in a metallic form, is so slightly attracted by a magnet that the pull can not be felt with the hand, whereas magnetic oxide of iron, containing about 70 per cent of iron in a nonmetallic form, is strongly attracted. A magnetometer or even a pocket compass needle will usually detect magnetism in commercial manganese steel, especially if a test be made soon after the steel has been in contact with a magnet, but the amount that may be so detected is extremely small. For most practical purposes it is zero.

Properties of Manganese Steel in the Raw State

The properties of manganese steel in the raw state are much like those of other raw high-carbon steels, the metal being very hard, but its ductility being practically negligible. The steel, because non-magnetic, may be used for purposes requiring a hard nonmagnetic metal, if it is not liable to shock. With better tool steels, which may make machining manganese steel a commercial operation, the metal may find a field of use in electrical apparatus, replacing some brass or other nonferrous metals because of its lack of magnetic qualities and lower cost.

HEAT TREATMENT OF MANGANESE STEEL

Although the composition of manganese steel is extremely important in determining its properties, the heat treatment to which it is subjected to develop in it its great toughness or ductility is even more so.

As used, it is almost universally water-toughened according to the method Hadfield set forth in his early papers* on the subject. This treatment consists in heating the whole article to about 1,050° C. and then cooling it as quickly as possible by immersing it in cold water, the colder the water and the more of it, the better. It will not do to heat only a part of the piece for quenching, and if a part of a toughened article becomes heated to redness or near it by accident or design the whole piece should be reheated and again quenched to give it proper qualities for use.

No time should be lost in completing the heating and quenching after the piece has become red-hot to avoid oxidation as completely as possible. Manganese steel is a poor conductor of heat, a factor that interferes with its heat treatment and tends to limit the thickness of the steel that may be profitably

* Hadfield, R. A., On manganese steel: Jour. Iron and Steel Inst., 1888, pt. 2, pp. 41–82; Manganese in its application to metallurgy: Proc. Inst. Civ. Eng., vol. 93, pt. 3, 1887-88, pp. 1–16; Some newly discovered properties of iron and manganese, loc. cit., pp. 61–126.

treated. This limit of thickness is generally taken as 4 inches, though somewhat thicker pieces in which the presence of internal cracks in the central parts would not be ruinous are treated in particular instances.

A thick piece of manganese steel must be heated at a slow rate. The heating can not be properly done by placing the piece when cold in a hot furnace. The furnace if hot from previous use must be cooled down nearly to atmospheric temperature or to that of the piece to be toughened, if it is not cold, before it is placed in the heating chamber. If the furnace is much hotter than the piece the poor heat conductivity of the metal retards the passage of heat to the interior, and the temperature of the skin rises rapidly, because of which and because of the high coefficient of expansion of the metal the exterior is expanded so much sooner than the interior that the latter is likely to be torn apart, the cracks extending outward from the center toward the surface until the stresses are sufficiently relieved or they reach metal too much softened by the heat to crack.

Many have asked when first considering the question, "Why not quench the casting when still hot with the initial casting heat?" The answer is that the heat of the various parts of the piece is not sufficiently uniform, the thicker parts being hotter than the thinner. If the piece is what might be called of uniform thickness, any corners or edges or parts first laid bare after casting are cooler than the rest of the piece, and even if the piece be a sphere, which is almost the only shape that will show a surface of uniform temperature after casting, the interior will be so much hotter than the exterior, and will, therefore, when the sphere is immersed in a cooling bath, continue contracting so long after the surface is cold and contraction there has ceased that it will tear itself apart. The toughness of the metal will not prevent internal rupture under such conditions as there can be no contraction of area, as in the pulling test, to compensate for and so permit stretching of the metal, which must therefore separate at its weaker spots and be subject to cracks.

PROPERTIES OF HEAT-TREATED MANGANESE STEEL

Hadfield's papers of 1888* gave the results of a large number of physical tests of manganese steels. These results indicate that the toughening effect of water quenching was imparted only when the steel contained at least 9 per cent of manganese. Since that date steels having as little as 7 per cent of manganese have been given a useful toughening effect though only a fraction of that of commercial manganese steel.

The hardness of toughened manganese steel is unique, and it may be termed a tough hardness and not a flinty hardness. Such steel may easily be dented with a hammer or marked with a file or chisel, but cutting it to a useful extent is almost impracticable, so that such finishing as is necessary is usually done by grinding with abrasive wheels.

The water toughening of manganese steel gives it great ductility—greater as to elongation, perhaps, than that of any other steel and exceeding sometimes 50 per cent in 8 inches, although its high degree of hardness is not greatly altered. This high ductility in combination with the great hardness of manganese steel gives it great resistance to abrasive wear as well as safety from breakage. Practically all manganese steel is used in the toughened state.

In the pulling test the percentage of contraction of area is less than the elongation, a result directly opposite to that with simple as well as most alloy steels, in which the percentage of contraction is usually twice or more that of the elongation. The pulled test piece has a rather uniform stretch throughout its length, whereas simple steels, as is well known, have a largely increased amount of stretch near the point of fracture. When a piece of manganese steel is pulled, the increase of strength due to cold working (stretching) is greater than the decrease in cross-section due to contraction, so that a stretched part becomes stronger than the unstretched parts, and elongation then occurs at another place. As the pulling is continued, all

* Hadfield, R. A., loc. cit.

parts of the pulled section stretch one after the other, with the result that when the piece is finally ruptured the stretch has been comparatively uniform. There is indeed an increased local extension and contraction close to the point of rupture, as with other ductile steels, but it is less marked in the manganese steels.

The elastic limit of manganese steel is unexpectedly low and not well defined, as the steel yields at a gradually increasing rate when pulled, as in testing, giving no point that strictly speaking, can be said to be the elastic limit or even yield point. The strain diagram has no jog.

A recent pulling test of forged, heat-treated, manganese steel gave the following results. The steel was cast in a test bar 3 inches square, forged down to a test piece of about the dimensions given, and finished by grinding.

RESULTS OF PULLING TEST OF PIECE OF MANGANESE STEEL

Diameter of piece, inches	0.823
Length, inches	2
Tensile strenght per square inch, pounds	152,840
Elastic limit per square inch, pounds	56,400
Elongation, per cent	51
Contraction, per cent	39.5
Carbon, per cent	1.10
Manganese, per cent	12.4
Silicon, per cent	0.15
Phosphorus, per cent	0.06

The length of the pulled section of a manganese-steel test piece does not affect the elongation as much as is the case with simple steels, because the stretch is so much more nearly uniform, as described above.

Owing to its lack of elastic limit and to its high ductility, manganese steel is prone to flow under stress, and it does not have high resistance to compression or to continually repeated blows of a hard mineral of other material that will gradually batter it out of shape. The results of tests of this property made with the Government testing machine at the Watertown Arsenal were as follows:

RESULTS OF COMPRESSION TESTS OF CAST MANGANESE STEEL

No. of Test Piece.	Analysis.				Permanent Set at a Pressure per Square Inch of				Total Load.
	C	Si	Mn	Cr	40,000 Pounds.	50,000 Pounds.	60,000 Pounds.	70,000 Pounds.	
	Per Ct.	Per Ct.	Per Ct.	Per Ct.	Inches.	Inches.	Inches.	Inches.	Pounds.
1	1.23	0.95	12.6	.....	0.0006	0.0036	0.0213	0.0981	190,100
2	1.26	.54	12.8	.....	.0020	.0046	.0182	.0899	180,100
3	1.31	.43	12.7	.....	.0010	.0036	.0204	.0998	172,300
4	1.22	.72	11.7	0.86	.0002	.0009	.0038	.0220	175,200

All test pieces were cast and finished by grinding to 4 inches long and 1.129 inches in diameter, giving 1 square inch of cross-sectional area. At the total load the pieces buckled. The permanent set at a pressure of 40,000 pounds per square inch shows that the limit of elasticity was passed in every case.

The hardness by Brinell's ball test of manganese steel is low, running usually about 190.

The merit number obtained by multiplying the figure representing the tensile strength of a material by the figure representing its elongation gives roughly a fair idea of the amount of work that must be expended upon the material to break it—that is, its strength must be overcome through a distance represented by its ductility before it will be broken. The merit number of manganese steel is perhaps the greatest of all known steels. The merit numbers of various metals are given below.

MERIT NUMBERS OF VARIOUS METALS

Metal.	Tensile. Strength.	Elongation.	Merit Number.
	Pounds.	Per Cent.	
Manganese steel	140,000	50	7,000,000
Soft steel	60,000	30	1,800,000
Tool steel.........................	130,000	5	650,000
Cast iron	20,000	.5	10,000
Nickel steel, natural................	95,000	21	1,995,000
Nickel steel, heat-treated	207,000	14	2,898,000

Of course, each of these metals yields to stress within its elastic limit, so that some work must be expended on it to strain it up to that point. With a nonductile metal, such as cast iron or even raw manganese steel, this is the reason for such small resistance to shock as it has. The elongation of 0.5 per cent given in the table is high for cast iron, but its power to absorb energy without breaking is increased by its ability to be strained within the elastic limit. If a material were to have no ductility and were perfectly rigid, the slightest blow would break it. Glass, having no ductility and being nearly rigid, comes near to being such a material. Of course, it may be easily broken by a blow, though it offers considerable resistance to a static load—enough to be useful, as in a floor plate, for example.

The effect of the toughening operation on the merit number of manganese steel is shown by the following results, taken from Hadfield's first paper.*

RESULTS OF TESTS OF TOUGHENING OPERATION ON MERIT NUMBER OF MANGANESE STEEL

Sample No.	Condition.	Tensile Strength.	Elongation.	Merit Number.
		Pounds.	Per Cent.	
1	As forged	88,120	3.5	308,420
	Toughened	145,240	50.0	[1] 7,262,000
2	As forged	81,600	1.56	127,296
	Toughened	150,370	44.44	[2] 6,682,440

[1] The toughened steel was 24 times as hard to break as the untoughened.
[2] The toughened steel was 50 times as hard to break as the untoughened.

MANGANESE-STEEL CASTINGS

Manganese-steel castings are made in dry sand, in green sand, and in some instances in iron molds, the considerations leading to the adoption of any particular material being much the same as with ordinary simple steel castings, such as danger of pulling apart or cracking in cooling, misrunning, or failure

* Hadfield, R. A., Manganese in iron and steel: Proc. Inst. Civ. Eng., vol. 93, pt. 3, 1887–88, pp. 1–16.

to fill the mold properly, and breaking or washing of the mold, and numerous others. The high coefficient of expansion of manganese steel must be considered, as it increases the liability of a casting to be cracked or pulled apart by shrinkage in cooling.

Manganese steel is prone to settle as it solidifies, demanding, for a given massiveness of design, larger sink heads than simple steels to feed the casting properly and prevent settle holes.

Even more than with other steel castings it is important that manganese-steel castings be so designed that the mass is fairly uniform throughout, or in particular that no part is much thicker than the rest. If a thick part is unavoidable, it should be connected with a sink head by metal as thick or nearly so. Thus, bosses and heavy fillets, often advisable in iron and simple steel castings, should be avoided because of the local increase of the mass they cause. The trouble is that a heavy part incompletely fed will be unsound in its central parts. A hole or recess cored in, if permissible, may prevent the central cavity, or an iron or soft steel core may be imbedded in the thick part, which, by hastening the solidification of the metal, may prevent the formation of holes or loose metal there.

USES OF MANGANESE STEEL

The specific uses for which manganese steel is employed may be learned from the catalogues of the producers. It is used in equipment employed in the mining, milling, and treatment of ores and other mineral products, quarrying and rock-dressing, digging and dredging, railway tracks, parts of machines exposed to gritty wear, and burglar-proof safes and vaults for the safe-keeping of money and valuables.

Because of its low-yield point manganese steel does not give satisfaction in many lines for which otherwise it seems to be eminently fitted.

When the service required is such that the resistance of the metal to flow is not exceeded, the results are often excellent. When for any reason the pressure at the point of wear passes the flowing point the results are disappointing, though even then they may be better than those obtained by the use of any other material.

In applying manganese steel to a new use determination of its fitness is nearly always a matter to be demonstrated, though this is less general than formerly as knowledge and experience have increased. Some of the fields that it was thought at first would be well occupied by manganese steel it has never won; others that were formerly unthought of it has found easy of conquest. Much uncertainty has come in applying it to new uses by the requirement that manganese steel as a metal be substituted for some other metal of a certain size and shape. The results are often not fully foreseen and liable to be unsatisfactory, as such a substitution is itself unscientific. The proper way to apply a new metal of unique properties to a new use is to study the service to be rendered and then to devise methods of doing the work and to design parts therefor such as will utilize to the best advantage the superior properties of the new metal. Sometimes, however, this is not easily done. Nevertheless manganese steel, because of its properties, presents difficulties that are usually best overcome by such a plan if the highest benefit attainable from its use is to be gained.

It is a well-established rule that manganese steel resists admirably abrasion under slow speeds of impact, as in Blake crushers, rolls, gyratory crushers, and similar machines, but results in high-speed grinders, such as the various centrifugal mills, are, if not poor, at least such as will not often warrant the expense of manganese-steel wearing parts, especially if such parts require some finishing, which must be done by slow and expensive grinding. In accordance with the same rules manganese steel makes good car wheels that are to run at slow speeds, such as on mine cars, but it is unsatisfactory in wheels used on railway cars that are run at high speeds.

For railway-track work manganese-steel cast frogs, switches, curved rails, and other special work are most excellent and they are extensively used.

The properties of the metal were early seen to make it an ideal material of which to make burglar-proof safes and vaults;

that is, it is too hard to be cut, and too strong to be broken even by considerable charges of dynamite and nitroglycerine. It has now been employed for this purpose some 15 years and no genuine manganese-steel safe has yet been opened by burglars, though a few have been attacked. Burglars have learned the futility of their efforts to rob such a safe.

The nonmagnetic property of manganese steel has found an important use in the cover plates of lifting magnets for handling heavy iron and steel articles where it is subjected to hard blows from the pieces jumping to meet the magnets. It offers little or no obstruction to the passage of the magnetic attraction. It is also used in the structure about the compasses on some ships because it does not affect the compass needle. This use may become important.

On the magnetic survey yacht *Carnegie* of the Carnegie Institution, which is built wholly of wood, bronze, and other nonmagnetic materials, manganese steel, because of its nonmagnetic property, is used in the grate and other parts of the gas producer which were necessarily made of steel.

HOT-WORKED MANGANESE STEEL

Manganese steel is, like simple steel, or even more so, improved in its physical properties by hot working (forging or rolling). Cast test pieces usually give misleading results because of imperfections due to casting. To cut out of the solid and finish by grinding a test piece of cast manganese steel is expensive, and when done the piece is liable to be so imperfect as to give practically valueless information. In this respect, however, it is like many other cast metals.

A steel that cast and heat-treated may show a tensile strength of 80,000 pounds per square inch with 20 per cent elongation may have, when well worked by forging and rolling and then heat-treated, a tensile strength of 140,000 pounds per square inch and 50 per cent of elongation in 8 inches.

The first commercial use of manganese steel was in dredge pins for chain bucket dredges, which were forged from square ingots in 1889 or 1890 by Hadfield at Sheffield. The greater ease of casting the metal, however, led to the employment of castings wherever they could be used. The records made by castings have stimulated in recent years the production of rolled pieces, particularly rails, plates for chute linings, and screens for crushed stone. A beginning has been made in the use of forged manganese steel for purposes for which its great strength and ductility (its merit number) and the consequent margin of safety it affords make it particularly reliable. Its use in spring hangers on locomotives is an example. Cold working such as stretching or cold rolling rapidly raises its tensile strength and elastic limit but destroys most of its ductility. Cold-rolled manganese steel on test has shown a tensile strength of 250,000 pounds per square inch and an elastic limit of 230,000 pounds.

For screening coke woven screens of manganese-steel bars are giving a promise of life a hundred times as great as that of screens made of the soft simple steel usually used for this purpose. Coke rapidly wears away metal on which it impinges, as though it were composed of or contained many minute diamonds.

The high electrical resistance of manganese steel would no doubt lead to its general use as a resistance material in the form of wire, for which it has been tried, were it not that its hardness makes drawing it into wire too costly an operation. Its electrical properties are good for this purpose, its specific resistance running from 65 to 75 microhms per cubic centimeter.

Agricultural implements made largely of rolled plates, such as shovels and hoes, offer a most inviting field for the use of manganese steel, especially for kinds of service in which the plates are used up by wear. If the service is such that they rust or corrode much more than they wear, tools made of simple steel may be as good. This field is now being entered.

The largest demand for hot-worked manganese steel is in rails for railroads. The rails are rolled on ordinary rail mills and are heat-treated by being quenched immediately after rolling. The service rendered by the rails is excellent and their use is extending. Some railroad men think their durability at least five times that of ordinary rails. Their value to the railroads is not, however, as great as their service compared with that of simple steel rails as the interest on the extra price and the relatively small value of the scrap when they are worn out tell against them, though the saving in the cost of laying is in their favor.

Perforated plates of manganese steel for screening ores, crushed stone, and other mineral products are promised, but special punching machinery is needed to punch the holes, as they probably must be punched at a much slower speed than that at which simple steels are punched.

CONCLUSIONS

The economies resulting from the use of manganese steel in its various fields of service have been great. Perhaps the most conspicuous example of its use was in connection with the Panama Canal, where years of time and millions of expense were saved by its use.

The promise of usefulness because of its great tensile strength and more remarkable ductility, there is ground to hope, may in the near future be fulfilled.

BIBLIOGRAPHY ON MANGANESE STEEL

HADFIELD, R. A. On manganese steel. Jour., Iron and Steel Inst., 1888, pt. 2, pp. 41–82.

—— Manganese in its application to metallurgy. Proc. Inst. Civ. Eng., vol. 93, pt. 3, 1887–88, pp. 1–60. Some newly discovered properties of iron and manganese, pp. 61–126.

—— Iron alloys with special reference to manganese steel. Proc. Am. Inst. Min. Eng., vol. 23, 1893, pp. 159–196.

MUKAI, TETSKICHI. Studien über chemisch-analytische und mikroscopische Untersuchung des Manganstahls. Friedberg, 1892.

HOWE, H. M. Note on manganese steel. Proc. Am. Inst. Min. Eng., vol. 21, 1892, pp. 625–631.

—— Manganese steel. Trans. Am. Soc. Mech. Eng., vol. 12, 1891, pp. 955–974.

—— The heat treatment of steel. Proc. Am. Inst. Min. Eng., vol. 23, 1893, pp. 467–476.

LEDEBUR, A. Ueber Manganstahl. Stahl und Eisen, Jahrg. 13, June, 1893, pp. 504–507.

HADFIELD, R. A. The results of heat treatment on manganese steel and their bearing upon carbon steel. Jour. Iron and Steel Inst., 1894, pt. 1, pp. 156–251.

GUILLET, LÉON. Aciers au manganèse. Soc. d'encourage. l'ind. nat., t. 102, 1903, pp. 421–448.

—— Recherches sur les aciers au manganèse. Génie civil, t. 43, 1903, pp. 261, 280.

—— Nouvelles recherches sur les aciers au manganèse. Rev. mét., t. 1, 1904, pp. 89–91.

LEDEBUR, A. L. Guillet's Untersuchungen über Manganstahl. Stahl und Eisen, Jahrg. 24, March 1, 1904, pp. 281–285.

ARNOLD, J. O., and KNOWLES, F. K. Preliminary note on the influence of manganese on iron. Jour. Iron and Steel Inst., 1906, pt. 1, pp. 106–124.

LAKE, E. F. Manganese steel and some of its uses. Am. Machinist, vol. 30, pt. 1, May 16, 1907, pp. 700–702.

GÉNIE CIVIL. L'acier au manganèse. T. 52, 1908, pp. 288–290.

RAILROAD GAZETTE. Rolled manganese steel rail. Vol. 45, December 11, 1908, pp. 1536–1538.

IRON AGE. Manganese-steel rails rolled at the Passaic works. Vol. 82, September 10, 1908, p. 709.

REVUE DE METALLURGIE. L'acier au manganèse. Vol. 6, May, 1909, pp. 551–651. Letter from Schneider & Co., in reply to note in the Iron Age of September 10, 1908.

BURGESS, C. F., and ASTON, JAMES. Observations upon the alloys of iron and manganese. Electrochem. and Met. Ind., vol. 7, November, 1909, pp. 476–478.

COE, H. I. Manganese and its effect on iron and steel. Proc. Staffordshire Iron and Steel Inst., vol. 25, pp. 16–47. Includes discussion and bibliography.

AMERICAN MACHINIST. A machine shop without cutting tools. Vol. 32, pt. 2, Nov. 25, 1909, pp. 893–897.

IRON AGE. Manganese-steel products. Vol. 84, Sept. 30, 1909, pp. 984–987.

POTTER, W. S. Manganese steel. Jour. West. Soc. Eng., vol. 14, 1909, pp. 212–240.

IRON AGE. Rolled manganese steel rails. Vol. 83, April 22, 1909, pp. 1261–1262.

STRONG, J. B. Rolled manganese-steel rails. Ry. and Eng. Rev., vol. 49, March 13, 1909, pp. 214–215.

ARNOLD, J. O. and READ, A. A. The chemical and mechanical relations of iron, manganese, and carbon. Jour. Iron and Steel Inst., 1910, pt. 1, pp. 169–184.

JOHNSON, F. E. Manganese steel. Jour. Assn. Eng. Soc., vol. 45, November, 1910, pp. 175–183.

LAKE, E. F. The various effects of manganese in steel. Am. Machinist, vol, 33, pt. 1, March 10, 1910, pp. 456–457.

SPRINGER, J. F. Manganese steel. Cassier's Mag., vol. 39, December, 1910, pp. 99–116.

KNEISLEY, G. W. The history and manufacture of manganese steel. Central and Eng. News, vol. 23, 1911, pp. 252–254.

HILPERT, S., and MATHESIUS, W. The magnetic properties of manganese and nickel steels. Jour. Iron and Steel Inst., 1912, pt. 2, pp. 302–310. Extract of a report published in Stahl und Eisen, 1912, p. 96, and also of the thesis of W. Mathesius, Berlin, 1912. Technische Hochschule.

FOUNDRY. Making manganese steel castings machineable. Vol. 40, July, 1912, p. 271. Describes method patented by F. Kohlhaas, Düsseldorf, Germany.

HADFIELD, R. A. Heating and cooling curves of manganese steel. Jour. Iron and Steel Inst., pt. 2, 1913, pp. 191–202.

HALL, J. H. Manganese-steel castings. Iron Age, vol. 91, March 20, 1913, pp. 712–713. Describes practice of the Taylor-Wharton Iron & Steel Co.

RAILWAY AND ENGINEERING REVIEW. New track appliances; manganese steel. Vol. 53, Oct. 11, 1913, pp. 955–956. From report to the Roadmasters' and Maintenance of Way Association, Chicago, September 11, 1913.

IRON TRADE REVIEW. Practice of the Edgar Allen American Manganese Steel Co. Vol. 52, June 19, 1913, pp. 1404–1411.

STONE, S. R. Manganese steel for machinery parts. Iron Age, vol. 91, Jan. 9, 1913, pp. 140–142.

ELECTRIC RAILWAY JOURNAL. Chicago's experience with solid and insert manganese special track work. Vol. 43, May 30, 1914, pp. 1221–1222.

HADFIELD, R. A., and HOPKINSON, B. The magnetic and mechanical properties of manganese steel. Jour. Iron and Steel Inst., pt. 1, 1914, pp. 106–137.

HADFIELD, R. A. Manganese-steel rails Bull. Am. Inst. Min. Eng. February, 1914, pp. 259–265.

HOPKINSON, B., and HADFIELD, R. A. Research with regard to the non-magnetic and magnetic conditions of manganese steel. Bull. Am. Inst. Min. Eng., March, 1914, pp. 513–530.

POTTER, W. S. Manganese steel, with especial reference to the relation of physical properties to microstructure and critical ranges. Bull. Am. Inst. Min. Eng., April, 1914, pp. 601–634.

ZERHANSEN F. R. How manganese-steel castings are made; details of molding, melting, and pattern making. Foundry, vol. 42, April, 1914, p. 132.

CHAPTER V

SIMPLE NICKEL STEELS

Nickel steel was chronologically the fourth alloy steel to be introduced, it having been in use for 25 years, and the steels to which nickel is added aggregate a large tonnage. In 1913, including the nickel-chromium steels as well as the simple nickel steels, the total amount of steel produced to which nickel was added was about 150,000 tons in the ladle, of which about 50,000 tons was simple nickel steel. The field for the latter is being steadily narrowed by the substitution of the cheaper or better nickel-chromium steels.

The useful nickel-iron alloys range, with large intervals, from 2 to 46 per cent of nickel, a greater compass than is covered by any other element alloyed with iron. The addition of less than 2 per cent of nickel alone does not seem to give enough benefit to make the addition worth while.

Nickel in untreated ordinary nickel steel raises the tensile strength, and in a greater proportion the elastic limit for a given content of carbon without decreasing the ductility.

Nickel steels with the different percentages of nickel present about the same range of internal microscopic structures as do manganese-iron alloys. With low-nickel content, as in the great bulk of nickel steels made, the unhardened steel is pearlitic. Higher nickel content gives martensitic structure and still higher austenitic. Certain steels with each of these structures find uses as is noted in subsequent pages

MANUFACTURE OF SIMPLE NICKEL STEEL

Nickel steel is made by any of the steel-making processes, but most of it is produced in the open-hearth furnace. The operations are similar to those followed in the production of simple steels, the nickel being either in the materials of the original charge or added in the metallic form at any time long enough before the heat is cast for the nickel to be melted and thoroughly mixed with the metal of the charge. Nickel is negative to iron at steel-melting temperatures and the iron protects it from oxidation and even reduces it from its oxide so that it is not wasted to any considerable extent in melting or working even when iron ore is added to the bath. On the other hand it does not deoxidize the metal or decompose carbonic oxide or keep the hydrogen and other gases in solution. It is not added, therefore, for curative purposes, as it gives no aid in rendering steel sound, or free from holes. In fact, nickel steel is prone to have seams and surface defects after it has been rolled, which is one reason against its wider use. The service of nickel is merely as an alloying element, to improve the physical properties of the finished steel either in its natural or heat-treated condition.

As might be expected from an alloying metal whose atomic weight, specific gravity, and fusion point are so near those of iron, nickel does not segregate much as the steel solidifies, and is claimed to hinder in some degree the segregation of carbon and the other less metallic ingredients. A reasonable explanation of this action of nickel on the other elements has not yet been established.

WORKING OF SIMPLE NICKEL STEEL

Ordinary simple nickel steel (3 to 4 per cent nickel) is worked hot by the usual forging and rolling operations much as simple steel is worked. The higher nickel steels are more difficult to work, having narrower ranges of temperature at which they may be hot-worked without showing signs of red-shortness. In the ordinary grades seams and adhering scale give some trouble.

Although molten iron protects molten nickel from oxidation, as stated, iron can not protect nickel from oxidation in scale formed on nickel steel, as in the heating furnace. The scale formed sticks much more firmly to the metal than that of simple steel, both hot and cold, and requires particular measures for its removal. Articles such as plates, having large flat surfaces from which the scale can not escape even if loosened by such means as rolls or the flat dies of the forging press, are sometimes cleaned cold by electrically driven machines which break up the scale by a shower of blows with chisels or hammers. A round article is much more satisfactorily cleaned of scale by the forging operation, as the metal is worked while not in contact with the dies, and the scale is thereby loosened and falls off. The scale naturally escapes more easily from the under side of a rolled plate than from the top in rolling, and some mills are equipped with apparatus for turning the plate bottom up before it is finished, so that the scale which has been repeatedly loosened and then rolled in again by the rolls can drop off when loosened from the under side.

Steels containing useful quantities of nickel are liable to contain seams that appear as dark-colored lines in the metal. The seams doubtless come, sometimes at least, from "skin" gas holes which become oxidized on their walls, and, although such oxidized holes will, if present, form seams in any steel, they seem to persist more in nickel steel because, perhaps, the nickel prevents the welding of such holes, as may happen with holes in simple steels if they are squeezed together while hot enough and the walls are clean and not oxidized. A hole near the surface in nickel steel might conceivably, therefore, be drawn out and the slit formed be opened to the air by the hot working, then oxidized on its inner surface, and form a seam, when a similar hole in simple steel would be welded and therefore not form a seam. Nickel-steel ingots should therefore be made sound and free from gas holes. It is held by some persons that seams develop in rolling without being caused by gas holes, and that this tendency is lessened by rolling at a high temperature, about 1,300° C. (2,372° F.).

GENERAL CHARACTER OF NICKEL STEEL USED FOR STRUCTURAL PURPOSES

The great bulk of simple nickel steels contain from 2 to 4 per cent of nickel, a proportion that affords the most suitable physical properties for nearly all structural purposes, and the nickel content usually aimed at in steels for structural purposes is 3.25 per cent. This grade might be called ordinary nickel steel as it is usually meant when nickel steel is mentioned without further specification. It has high value for structural purposes such as bridges, gun forgings, machine parts, engine and automobile parts, and any similar line of service that is too severe for simple steels.

The bridges in which it is used are particularly those of great span, and it is nearly always used in the natural or annealed condition when the additional strength and ductility imparted is that due to the mere presence of nickel in the metal. Important quantities are used in the Queensboro, Manhattan, St. Louis Municipal, and Quebec bridges, and the Kansas City viaduct. Some nickel-steel tension bridge members have been heat treated by heating and quenching, being immersed in water edgewise with the longitudinal axis horizontal, and afterward drawn back by a second heating to give an elastic limit of 55,000 pounds per square inch, a rather low figure. A relatively low percentage of nickel, or about 2 per cent, is sufficient to afford steel with such a property, when heat treated.

The use of nickel steel in bridges saves some weight, a detail of importance in such bridges as those mentioned, but when the span is moderate a bridge of simple steel is perhaps as good and is less costly even though it contains considerably more weight of metal.

Steel with 2 per cent of nickel is used in seamless tubes such as are used for bicycles and for other equipment requiring a high-grade tube. They are not heat treated, but higher properties than those of the steel in its natural state are imparted by the cold-drawing operations by which these tubes are finished. The ordinary grade with 3.5 per cent nickel is used in cannon, being always heat treated for this use. It is also used in many automobile parts, the variety of high properties obtainable in it by modifying its heat treatment rendering it fit for almost any service demanding a strength and security from breakage that a simple steel will not meet.

In some large dynamos the revolving fields are connected by nickel-steel rings having 3 per cent nickel, the metal being particularly well suited for the purpose both by its strength and its magnetic efficiency, the permeability being high and the hysteresis losses low.

CHARACTERISTICS OF DIFFERENT NICKEL STEELS USED IN RAILS

Nickel-steel rails usually having about 3.5 per cent of nickel have been tried by many railroads and are generally considered unsatisfactory, though small lots are still being made chiefly for use in tunnels and other unusually wet or damp places, both for their ability to resist rusting and for safety from breakage. Their price is nearly twice that of simple steel rails and they sometimes give three times the service, but their average life has been much less than that. One lot of Bessemer nickel-steel rails which gave between one to two times the service of simple Bessemer steel rails wore unevenly, and the metal flowed over the side of the head on the curve so that it was finally detached in thin splinters, some of which were 3 or 4 feet long. The rails from the upper parts of the ingots were more unsatisfactory than those from the bottom, and though the ingots were not examined for their soundness it seems evident that their upper parts were infested with blowholes as well as pipe, and that none of the holes or pipe were welded up in rolling, as the effect of nickel in hindering welding of steel is well established.

Properties of Ordinary Nickel Steel

The properties of ordinary nickel steel are given below. All the samples consisted of small test pieces, and elongations were measured on 2 inches except as noted.

PROPERTIES OF ORDINARY NICKEL STEELS

Sample No.	Composition.						Condition.	Physical Properties.				
	C	Mn	Si	S	P	Ni		Tensile Strength.	Elastic Limit.	Elongation.	Contraction.	Ball Hardness.
	Per Cent.	Per Cent.	Per Cent.	Per Cent.	Per Cent.	Per Cent.		Pounds.	Pounds.	Per Cent.	P r Cent.	
1 [1]	0.28	0.57	....	0.03	0.02	3.44	Natural state..	95,420	56,670	[2] 21.2	50	....
2 [3]	.40	.64	....	.02	.01	3.43	Annealed.....	98,800	51,400	[4] 12.4	33	....
3 [3]	.40	.55	....	.03	.01	3.70	Annealed.....	93,180	56,060	[4] 15.8	40	....
4 [5]	.20	.65	....	.04	.04	3.5	Annealed.....	......	43,000	27	62	170
5 [5]	.20	.65	....	.04	.04	3.5	[6]	......	95,000	20	72	216
6 [5]	.20	.65	....	.04	.04	3.5	[7]	......	140,000	14	61	330
7 [5]	.30	.65	....	.04	.04	3.5	Annealed.....	......	63,000	27	63	163
8 [5]	.30	.65	....	.04	.04	3.5	[8]	......	87,000	25	68	207
9 [5]	.30	.65	....	.04	.04	3.5	[9]	......	123,000	15	57	269
10 [5]	.30	.65	....	.04	.04	3.5	[10]	......	187,000	13	57	405
11 [11]	.25	.74	0.21	.01	.01	3.55	[12]	207,000	177,000	14	60	395
12 [11]	.25	.74	.21	.01	.01	3.55	[13]	135,000	117,000	20	67	267

[1] Sample represented untreated steel for Quebec bridge.
[2] In 8 inches.
[3] Full size eyebars for St. Louis Municipal Bridge.
[4] In 18 feet.
[5] Figures taken from Fourth report of Iron and Steel Division: Bull. Soc. Automobile Eng., vol. 4, 1913, p. 168.
[6] Quenched in water at 850° C.; hardness drawn in air at 538 C.
[7] Quenched in water at 800° C.; hardness drawn in air at 316° C.
[8] Quenched in water at 800° C.; hardness drawn in air at 593° C.
[9] Quenched in water at 800° C.; hardness drawn in air at 399° C.
[10] Quenched in water at 800° C.; hardness drawn in air at 316° C.
[11] Figures furnished by Halcomb Steel Co.
[12] Quenched in water at 843° C.; hardness drawn in air at 316° C.
[13] Quenched in water at 843° C.; hardness drawn in air at 538° C.

One per cent of nickel in ordinary nickel steel in the natural state raises the tensility about 6,000 to 8,000 pounds per square inch.

The table shows that ordinary nickel steels may be so made as to have a wide range of properties that make them suitable for any structural purposes for which they are not too expensive.

The properties of one grade of nickel-steel castings made for special purposes are as follows: Composition, C 0.20 per cent, Mn 0.50 per cent, Si 0.35 per cent, Ni 2.50 per cent; tensile strength, 85,000 pounds per square inch; elongation, 25 per cent; contraction, 40 per cent. This steel was not given treatment involving quenching, but was merely annealed.

Steel containing 5 to 8 per cent of nickel presents a sort of critical point, that proportion being the lowest at which, with the usual range of carbon, the structure is all martensitic and consequently very hard, the martensitic state being equivalent to the hardened state of simple steels. Such steel is difficult to work hot or cold, but can be rolled if proper care is used. It finds some usefulness in places where great resistance to shock is required, particularly in thin shield plates about 0.15 inch thick which are used on one side of the caisson of field artillery to protect the ammunition, and the men who serve it, from rifle fire. A sample analyzed for "Tests of Metals"* gave the following composition in percentages: C 0.42, Mn 0.49, Si 0.26, S 0.02, P 0.02, Ni 6.68.

The content of carbon determines the minimum amount of nickel which must be present to make the steel wholly martensitic. Thus if the carbon content is low, about 0.20 per cent, 8 per cent of nickel is required, whereas if the carbon content is about 0.80 per cent the steel is martenistic when there is 5 per cent of nickel contained. The analysis given last above represents martensitic steel.

Guillet† gives the properties of a similar steel, with 6 per cent of nickel and 0.38 per cent carbon, as follows:

* Report of the tests of metals and other materials for industrial purposes, 1907, War Department, 1908, p, 42.

† Guillet, Léon, Nouveaux essais au choc a témperature variables: Rev. mét. t. 7, October, 1910, pp. 837–844.

PROPERTIES OF NICKEL STEEL CONTAINING 6 PER CENT OF NICKEL

Condition.	C	Ni	Tensile Strength.	Elastic Limit.	Elongation.	Contraction.	Shock.
	Per Cent	Per Cent	Pounds	Pounds	Per Cent	Per Cent	
Annealed	0.38	6.0	113,760	99,540	20	65	30
Air hardened at 850° C.........	.....	.....	177,750	156,420	11	53	19
Quenched in water	.....	.....	199,080	177,750	10	50	17

He does not say whether this steel was martensitic, but the high elastic limits indicate that it was probably largely so, even in the annealed condition.

Steel with 8 per cent nickel has one transformation point at 510° C. (950° F.) where points Ar_1, Ar_2, and Ar_3 are all merged into one. Eight per cent is the highest useful content of nickel in nickel steel that is amenable to ordinary annealing and quenching operations. Hardening by quenching does not occur in steels containing 10 per cent or more of nickel which are on the contrary softened by heating and quenching.

NICKEL-IRON ALLOY DISCOVERED BY ARNOLD AND READ

The 13 per cent nickel-iron alloy with 0.55 per cent carbon discovered recently by Arnold and Read* is noteworthy as it seems to possess the highest strength of any of the nickel steels. It is so hard as to be unmachinable and the investigators mentioned were not able to drill it even to get some drillings for analysis, the composition mentioned being what they aimed at when making the steel. It has a yield point of about 134,000 pounds per square inch, a tensile strength of about 195,000 pounds, with 12 per cent of elongation in 2 inches. This gives a merit figure of about 2,300,000, which is very high for such a hard steel, though it does not compare with the 7,000,000 of forged manganese steel. Steel of this composition might have

* Arnold, J. O., and Read, A. A., The chemical and mechanical relations of iron, tungsten, and carbon, and of nickel, iron, and carbon: Proc. Inst. Mech. Eng., March–May, 1914, pp. 223–279.

been expected to show maximum strength as a result of Hadfield's experiments,* though he did not include this grade in his series of samples. He found that low-carbon steels with 11.4 and 15.5 per cent of nickel each had a tensility of 210,560 pounds, which was more than was possessed by the steels next above and below. The curve therefore should have reached a maximum between them with a nickel content of about 13.5 per cent.

Arnold and Read's steel should, of course, have a higher tensility, or about 215,000 pounds, to harmonize with Hadfield's, and further tests are needed to establish the exact path of the curve. Arnold and Read note that the composition of this steel nearly corresponds with the formula Fe_7Ni. With such properties as it possesses this steel is likely to find at least a limited field of usefulness.

PROPERTIES OF DIFFERENT STEELS HAVING DIFFERENT PERCENTAGES OF NICKEL

Before Arnold and Read's discovery of the 13 per cent grade, 15 per cent nickel steel was thought to have the greatest strength of all the nickel steels—that is, in the natural state. This variety has been employed in a few instances for shafting and similar service for which other steels failed, but the amount of it used is negligible in statistics. It is hard to machine, and heating followed by slow cooling does not soften it, though heating and quenching does enough to allow it to be machined slowly. It has a tensility of about 170,000 pounds and an elastic limit of 150,000 pounds per square inch, according to one observer, though, as stated above, Hadfield obtained 210,560 pounds tensility, with little ductility. It is likely that the properties desired when this steel was used, particularly its ductility, could now be surpassed by the much cheaper heat-treated ordinary nickel or nickel-chromium steels.

Eighteen per cent nickel-iron alloy, although not useful, is

* Hadfield, R. A., Alloys of iron and nickel: Proc. Inst. Civ. Eng., vol. 138, 1899, pp. 1–124.

worthy of note here because of its anomalous action (according to Sexton and Primrose*) when cooled from 200° C. (392° F.). At first it contracts uniformly until its temperature falls to 130° C. (266° F.). Then it expands while cooling to 60° C. (140° F.), when it again contracts as the temperature falls.

Twenty-two per cent nickel steel is used when resistance to rusting or corrosion is desired. A noted example is the valve stems of the salt-water fire-protection service of the city of New York where the apparatus may not be allowed to become inoperative or hard of action from the formation of rust. It is also used sometimes for the spark poles in the spark plugs of internal-combustion engines, including automobiles, though commercial nickel wire is more commonly used.

High-nickel steels having 25 per cent or more of nickel and low carbon content (about 3 per cent) are austenitic in structure and in the natural state are softer and tougher than the medium-nickel martenistic steels.

High-nickel steel containing 24 to 32 per cent nickel in the form of wire is used for electrical resistance in small quantity, probably between 5 and 10 tons per year in this country.

The analysis and resistance of samples of Krupp nickel-steel resistance wire are shown below. This wire is used in electric toasters, cookers, irons, and similar devices.

ANALYSIS AND RESISTANCE OF SAMPLES OF KRUPP NICKEL-STEEL RESISTANCE WIRE

Sample No.	C	Mn	Si	S	P	Cr	Ni	Specific Resistance per Cubic Centimeter.
	Per Cent	Per Cent	Per Cent	Per Cent	Per Cent	Per Cent	Per Cent	Microhms.
1	0.52	0.75	0.10	0.035	0.024	.....	30.6	87.9
2	.39	1.00	.70	.035	.025	.....	24.2	....

Steel with 27 per cent of nickel is used in bits, stirrups, and spurs in riding harness because of its resistance to rusting. It

* Sexton, A. H., and Primrose, J. S., The metallurgy of iron

will nevertheless rust slowly at ordinary temperature under conditions that strongly induce oxidation.

Steels containing more than 24 per cent of nickel are practically nonmagnetic in their ordinary condition, a rather remarkable fact when the high magnetic susceptibility of both iron and nickel alone is considered. The explanation that the critical point marking the change from the nonmagnetic to the magnetic state of iron is lowered by the nickel from about 700° C. (1,292° F.) to below ordinary atmospheric temperatures is, no doubt, sound as far as it goes. When 25 per cent nickel steel is cooled to −40° C. (−40° F.) it becomes magnetic, and retains its magnetism at ordinary atmospheric temperatures. On being heated to 580° C. (1,076° F.), however, the alloy reverts to the nonmagnetic state. This separation of 620° C. between the critical points marking the magnetic states in heating and cooling is great in comparison with the 25° to 50° C. of simple steels, and because of it this steel is classed as irreversible.

The nonmagnetic quality of high-nickel steels is not utilized chiefly because of its capacity for becoming magnetic. as described above, for if it happened to be cooled enough to make it magnetic it could not in most cases be easily demagnetized.

The fact that high-nickel austenitic steels have a somewhat lower modulus of elasticity than the low-nickel or simple steels does not affect their value for the uses made of them. These steels also have low elastic limits, though they are tough and show up well in the shock test. Nevertheless they are generally used not because of superior physical properties, but because of their resistance to rusting and corrosion or their electrical resistance. With a carbon content of 0.25 to 0.30 per cent and 32 per cent nickel they are used in valves for gasoline motors with good results.

Nickel steel with 30 per cent of nickel is used in boiler tubes, particularly in marine boilers, for which it is admirable. These tubes are in the natural, not heat-treated state. They resist corrosion better than simple steel tubes and last three times as long. Hence their use is sometimes economical in spite of the much higher cost.

INVAR

The 36 per cent nickel steel known as Invar is used to the extent of perhaps a few hundred pounds a year in clock pendulums, rods for measuring instruments, and such parts for which its exceedingly slight expansion and contraction when heated and cooled within the atmospheric range gives it a particular value. Nevertheless, its coefficient of expansion, even though small, is not negligible, and compensating means must be used in Invar clock pendulums and in the Invar balance-wheels of watches. A watch with an Invar balance-wheel varied 20 seconds per day during a temperature change of 40° to 90° F., the usual test change, a variation too great for a good watch. Some Invar has as low a coefficient of expansion as 0.0000008 per degree C., and samples have been made that contracted slightly when warmed. The coefficient given indicates an expansion of 0.05 inch in a mile per degree C.

When Invar is heated to about 300° C. (572° F.) and higher its coefficient of expansion is greatly increased and its lack of expansion at ordinary temperatures appears to be merely a belated and not destroyed function. With excessive cold there is likewise a resumption of contraction.

PLATINITE

Forty-six per cent nickel steel with 0.15 per cent carbon known as platinite, has about the same coefficient of expansion as platinum and glass and for that reason may be imbedded in glass without breaking the latter by a difference in expansion. It has been used in leading wires in the glass bases of electric incandescent lamp bulbs as a substitute for platinum, which was formerly held to be indispensable. In those lamp bulbs the preservation of the vacuum is necessary and the joint between the wire and glass must be kept tight. Platinite has not been found wholly suitable for this purpose and is not now so used, a compound wire with a 38 per cent nickel-steel core encased in copper and sometimes coated with platinum being now generally employed. The nickel-steel core, if free, will expand less than the glass and the copper more, so that each resists the other, and the wire as a whole will have the desired rate of expansion. About 2 tons of nickel steel per year is used in this wire.

Many other alloys of iron and nickel have been studied by Guillet and others.* In fact the whole range has received more or less thorough attention, and much knowledge of scientific value has been gained concerning the varieties that so far have not found useful application.

BIBLIOGRAPHY ON NICKEL STEEL

BROWNE, D. H. Nickel steel; a synopsis of experiment and opinion. Trans. Am. Inst. Min. Eng., vol. 29, 1899, pp. 569–648. Includes a bibliography dating from 1889.

SOUTHER, HENRY. Nickel and iron alloys. Mineral Ind., vol. 8, 1899, pp. 438–439.. Briefly discusses the uses of nickel steel.

DUMAS, L. Recherches sur les aciers au nickel a haute teneurs. Ann. des mines, t. 1, sér. 10, 1902, pp. 357–561. Includes a bibliography.

PORTER, H. F. J. Nickel steel; its practical development in the United States. Cassier's Mag., vol. 22, 1902, pp. 480–500.

COLBY, A. I. A comparison of certain physical properties of nickel steel and carbon steel. 1903. Contains numerous bibliographies.

—— Nickel steel; its properties and applications. Proc. Am. Soc. Test. Mat., vol. 3, 1903, pp. 141–168.

—— Noncorrosive nickel-steel boiler tubes. Trans. Soc. Naval Arch. and Marine Eng., vol. 11, 1903, pp. 115–137.

WAR DEPARTMENT. Report of the tests of metals and other materials for industrial purposes, 1903. 1904, pp. 167–309. Discusses carbon and nickel-steel ingots.

GUILLAUME, C. E. Les applicationes des aciers au nickel. Paris, 1904.

JOHNS, COSMO. Notes on the production and thermal treatment of steel in large masses. Jour. Iron and Steel Inst., 1904, pt. 1, pp. 61–97. One table gives results of tests of hollow forged nickel steel forgings.

CARPENTER, H. C. H., HADFIELD, R. A., and LONGMUIR, PERCY. Seventh report to the Alloys Research Committee; on the properties of a series of iron-nickel-manganese-carbon alloys. Proc. Inst. Mech. Eng., 1905, pts. 3 and 4, pp. 857–1041.

DIXON, EDWARD. Nickel and carbon steels. Engineering, vol. 82, July 6, 1906, pp. 22–23.

* See bibliography following.

WATERHOUSE, G. B. The influence of nickel and carbon on iron. Electrochem. and Met. Ind., vol. 4, 1906, pp. 451, 482.

—— Nickel steel and its application to boiler construction. Iron Age, vol. 77, Feb. 8, 1906, pp. 490–491.

LAKE, E. F. The forging of alloy steels. Am. Machinist, vol. 30, pt. 2, Aug. 29, 1907, pp. 289–291.

GREGORETTI, U. F. La Corazza " Krupp cementata " esaminata al microscopio. Rivista Marittima, vol. 41, November, 1908, pp. 227–238. Translation in Rev. mét., t. 7, 1910, pp. 260–285. Describes Krupp process of armor-plate manufacture at the Terni Steel Works.

PREUSS, ERNST. The strength of nickel-steel riveted joints. Carnegie Scholarship Mem., Iron and Steel Inst., vol. 1, 1909, pp. 60–142. Includes a bibliography.

PREUSS, ERNST. Zur Kenntnis der Festigkeitseigenschaften des Nickelstahles. Stahl und Eisen, Jahrg. 29, Mar. 24, 1909, pp. 422–425.

WEBSTER, W. R. Nickel-steel eyebars for Blackwell's Island Bridge. Trans. Am. Soc. Civ. Eng., vol. 64, 1909, pp. 289–302.

WADDELL, J. A. L. Nickel steel for bridges. Trans. Am. Soc. Civ. Eng., vol. 63, 1909, pp. 101–387.

GUILLET, LÉON, and RÉVILLON, LOUIS. Nouveaux essais au choc a témperatures variables. Rev. mét., t. 7, October, 1910, pp. 837–844.

BOHNY, F. Ueber die Verwendung von Nickelstahl im Brückenbau. Stahl und Eisen, Jahrg. 31, 1911, pp. 89–97, 184–193. Paper read before the Verein Deutscher Eisenhüttenleute.

MCWILLIAM, ANDREW, and BARNES, E. J. Some properties of heat-treated 3 per cent nickel steels. Jour. Iron and Steel Inst., 1911, pt. 1, pp. 269–293.

COLVER-GLAUERT, E., and HILPERT, S. The thermal magnetic transformations of 25 per cent nickel steel. Jour. Iron and Steel Inst., 1912, pt. 2, pp. 295–301.

CONE, E. F. Properties of nickel cast steel. Iron Age, vol. 90, Aug. 8, 1912, pp. 287–288.

HODGE, H. W. Tests of two large nickel-steel columns. Eng. Rec., vol. 67, March 1, 1913, pp. 234–235.

ARNOLD, J. O., and READ, A. A. The chemical and mechanical relations of iron, tungsten, and carbon, and of iron, nickel, and carbon. Proc. Inst. Mech. Eng., March–May, 1914, pp. 223–279.

ENGINEERING RECORD. Ultimate strength of carbon and nickel-steel models of Quebec Bridge members. Vol. 69, Mar. 21, 1914, pp. 333–336; vol. 70, Nov. 14, 1914, pp. 541–543.

CHAPTER VI

NICKEL-CHROMIUM STEELS

NICKEL-CHROMIUM steels, known in the trade as chrome-nickel steels, are perhaps the most important of the structural alloy steels. Their field of usefulness is continually being enlarged by their application for new purposes and also by encroachment on the premises of some of the other alloy steels, notably of simple nickel steel, and they have almost wholly displaced nickel-vanadium and nickel-chromium-vanadium steels, which several years ago were in some considerable demand.

The amount of nickel-chromium steels produced in 1913 was thought to be about 100,000 tons of ingots, all made in the open-hearth furnace with the exception of 2,000 or 3,000 tons melted in crucibles and electric furnaces. The steel is made by 10 or 12 companies, 2 of which make it at several different plants.

Nickel-chromium steels are seldom used in any but a heat-treated condition. By suitable treatment pieces of small mass can be made to have as high physical properties as any steels known, with any elastic limit between 40,000 and 250,000 pounds per square inch, accompanied by ductility that is high as compared with its strength, as the ductility naturally lessens as the elastic limit increases.

Nickel-chromium steels can be made somewhat more cheaply than simple nickel steel of the same strength and ductility containing a smaller total of the alloying elements, and chromium is less costly than nickel.

COMPOSITION AND PROPERTIES

The upper limit of nickel in useful chrome-nickel steels is about 3.5 per cent, and all useful steels of this class are pearlitic according to Guillet.* According to the same authority, when a chrome-nickel steel is casehardened, the case is harder than that of a simple nickel steel.

Some of the defects and troubles of chrome-nickel steels are like those of simple nickel steels previously considered.

The composition and properties of six nickel-chromium steels in the natural or untreated state are given in the table following:

COMPOSITION AND PROPERTIES OF NICKEL-CHROMIUM STEELS IN NATURAL OR UNTREATED STATE

Sample.	Composition.							Tensile Properties.					Remarks.
	C	Mn	Si	S	P	Ni	Cr	Tensile Strength	Elastic Limit.	Contraction of Area.	Elongation in 2 Inches.	Ball Hardness.	
	%	%	%	%	%	%	%	Pounds	Pounds	%	%		
1	0.55	0.41	0.22	0.03	0.02	1.53	1.14	96,000	75,000	66	31	185	Annealed
2	.18	.27	.05	.04	.02	1.28	1.59	72,000	51,000	71	37	134	Annealed
3	.15	.34	.13	.02	.01	1.28	.37	59,000	42,000	64	38	115	Annealed
4	.29	.42	.07	.06	.02	3.86	1.48	.....	.....	....	....	....	Natural
5	.25	.32	.10	.03	.02	1.45	1.20	96,500	81,500	68	25	....	Test piece
6	.25	.32	.10	.03	.02	1.45	1.20	97,100	80,900	49	[1] 7	....	Eyebar; full size

[1] In 21 feet.

Sample 4 is from a plate similar to that used in the mast of the yacht *Vanitie*. It was not heat treated, but was used as rolled.

Samples 5 and 6 represent the same steel and show the relative properties of the small test piece and the full-size eyebar for a bridge the section of which was 14 by 2 inches. The difference in elongation is particularly noticeable, the great local stretch near the point of rupture being only a small part of the total length of the bar.

The composition and properties of six nickel chromium steels in the heat-treated condition were as follows:

* Guillet, Léon, Aciers nickel chrome: Rev. de mét., t. 3, 1906, pp. 462–484.

COMPOSITION AND PROPERTIES OF NICKEL-CHROMIUM STEELS IN HEAT-TREATED CONDITION

Sample No.	Composition.							Tensile Properties.						Heat Treatment.	
	C	Mn	Si	S	P	Ni	Cr	Tensility.	Elastic Limit.	Contraction of Area.	Elongation in 2 Ins.	Ball Hardness.	Temperature at which Steel was quenched in Water.	Temperature at which Temper was Drawn in Air.
	%	%	%	%	%	%	%	Pounds	Pounds	%	%		°C.	°C.
1	0.40	0.74	0.24	0.03	0.02	3.45	1.20	187,000	175,000	43	10	352	830	371
2	.36	.53	.11	.04	.01	1.55	.70	145,000	125,000	65	20	233	830	566
3	.21	.41	.22	.03	.02	3.52	1.11	110,000	75,000	66	24	215	830	682
4	.48	.44	.16	.01	.01	2.02	.98	212,000	186,000	46	10	445	843	427
5	.48	.44	.16	.01	.01	2.02	.98	140,000	120,000	61	18	287	843	649
6	.38	.28	.27	.02	.01	3.01	.65	114,000	90,000	69	25	266	843	649

Any one of the first three samples could be given substantially the properties of either of the other two by varying the temperature of the second heating.

Most of the nickel-chromium steel goes into armor plate, projectiles, and automobile parts.

USE IN AUTOMOBILES

For automobiles—and the practice might be advantageously extended to other fields—three grades of nickel-chromium steel are used. They are called low, medium, or high according to their contents of nickel and chromium. The carbon content may be varied for each grade within the limits shown in the following table:

COMPOSITION OF NICKEL-CHROMIUM AUTOMOBILE STEELS

Grade.	C	Mn	Si	S	P	Ni	Cr
Low...............	0.20 to 0.40	0.65	Low	0.045	0.04	1.25	0.6
Medium	.20 to .40	.65	Low	.045	.04	1.75	1.10
High	.20 to .40	.65	Low	.045	.04	3.50	1.50

These steels are almost invariably heat treated for use in automobiles, a wide range of properties being attainable by varying the heat treatment with each steel. The properties overlap those of steels of both harder and softer grades, so that a wide choice of properties is afforded as well as a choice of steels for the set of properties desired.

USE IN ARMOR PLATE

An important use for chrome-nickel steel is in both thick and medium armor plate for war vessels. The thick heavy side armor, 6 to 14 inches thick, is face hardened by the well-known methods. A recent analysis of the body of a plate gave: C 0.33 per cent, Mn 0.32 per cent, Si 0.06 per cent, S 0.03 per cent, P 0.014 per cent, Ni 4 per cent, Cr 2 per cent, and its tensile properties after treatment were:

Tensile strength, pounds per square inch	101,000
Elastic limit, pounds per square inch	77,500
Elongation in 2 inches, per cent	24
Contraction of area, per cent	60

The results from such a great mass of metal were excellent.

Medium armor, between 3 to 5 inches thick, is rather similar in composition. It is not face hardened, but is given high properties as a whole by the heat treatment to which it is subjected. An analysis lately made gave: C 0.30 per cent, Mn 0.34 per cent, Si 0.13 per cent, S 0.03 per cent, P 0.03 per cent, Ni 3.66 per cent, Cr 1.45 per cent.

Its physical properties were those given below as sample 1. Sample 2 represented another chrome-nickel steel made for the same purpose, containing $3\frac{1}{2}$ per cent of nickel.

	Sample 1.	Sample 2.
Tensile strength, pounds per square inch	119,000	138,000
Elastic limit, pounds per square inch	106,000	119,000
Elongation in 2 inches, per cent	22	22
Contraction of area, per cent	61	49

Such steel is most excellent for use on warships to form protective decks and barriers to protect from secondary battery fire. Chrome-nickel-vanadium steel is also used for this purpose, as noted elsewhere.

USE IN PROJECTILES AND IN RAILS

Nickel-chromium steel is used in the manufacture of most armor-piercing projectiles.

Cubillo* cites a steel for projectiles, having 0.48 per cent C, 0.58 per cent Mn, 0.75 per cent Cr, 2.55 per cent Ni, 0.40 per cent Si, 0.04 per cent P. A test piece quenched in oil and tempered had an elastic limit of 129,400 pounds per square inch, a tensile strength of 150,300 pounds per square inch, and an elongation of 19 per cent.

For large projectiles Girod† prefers chromium-tungsten steel having 0.50 per cent C, 4 per cent Ni, 0 to 1.5 per cent Cr, and 0.25 per cent W.

It is curious that nickel is considered to improve the quality of shot, although generally held to injure the quality of high-speed tool steels. In use there seems to be a parallel between the requirements of the two, except for the important and vital difference as to the required speed at which they respectively meet the metal to be penetrated. The speed of impact of the shot enables it to enter when no amount of pressure will effect the same result.

Chrome-nickel steel rails having 2 per cent of nickel and 0.7 per cent of chromium have been tried by several railroads, but with unsatisfactory results. They resisted wear well as compared with simple steel rails, but broke badly both transversely and lengthwise, so that they were considered unsafe and consequently were removed. They were made by the Bessemer process and were not heat treated.

DETAILS OF MANUFACTURE OF A SPECIFIC PIECE OF NICKEL-CHROMIUM STEEL

Following is a description of the manufacture of a large shaft of mild chrome-nickel steel for marine purposes. A cor-

* Cubillo, L., Armor-piercing projectiles: Jour. Iron and Steel Inst., 1913, p. 251.

† Girod, P., Discussion of paper on armor-piercing projectiles: Jour. Iron and Steel Inst., 1913, p. 252.

rugated 35-ton ingot 45 inches in diameter was made of basic open-hearth steel having 0.24 per cent C, 0.70 per cent Mn, 0.013 per cent P, 0.015 per cent S, 0.18 per cent Si, 1.60 per cent Ni, and 0.32 per cent Cr. A few hundredths per cent of titanium was added in the ladle, but did not appear in the steel. The shaft when finished was $14\frac{1}{2}$ inches in diameter, with an 8-inch hole through on the center line.

The steel was melted without the addition of ore late in the heat, a method that favored soundness and tended to allow the steel to clean itself of insoluble impurities such as oxides and silicates. The ingot was forged, annealed at 866° C. (1590° F.), bored, rough-turned, heated to 750° C. (1382° F.), quenched in oil, and drawn at 593° C. (1100° F.).

The shaft was merely raised to the drawing temperature, 593° C., when firing at once ceased, the furnace was closed, and the shaft allowed to cool with the furnace.

The averages of the tests, which were longitudinal, were as follows: Tensile strength, 83,300 pounds per square inch, elastic limit, 52,500 pounds per square inch, elongation in 2 inches 26 per cent, contraction 60 per cent. The results were excellent, though seemingly a lower drawing temperature, which would have resulted in a higher elastic limit, would have been justified.

MAYARI STEEL

A so-called natural chrome-nickel steel is now made from certain ores mined at Mayari, Cuba. The ores carry enough nickel to give 1.3 to 1.5 per cent of nickel, and enough chromium to give $2\frac{1}{2}$ to 3 per cent of chromium in the crude iron smelted therefrom. When the iron is converted into steel by the pneumatic or open-hearth processes, the nickel is practically all present in the steel, but the chromium is of necessity largely wasted by being oxidized.

Steel made in part of Mayari iron is giving good service in rails and particularly in track bolts, which are heat treated to give the metal an elastic limit of 75,000 pounds per square inch.

Why these rails are satisfactory when other chrome-nickel

steels were not has not been explained. The chief differences seem to be (1) that these Mayari steel rails have less of the alloying elements because Mayari iron is used only in part in them, and (2) that the steel is made in the open-hearth furnace.

The use of steel containing Mayari iron is increasing, and the demand is enough to induce the production synthetically of steels of the same composition by adding nickel and chromium to simple steels in the Mayari proportions.

The Mayari steels are not included in the estimated quantity of the chrome-nickel steels made, as already given. In fact it is likely that in the near future the tonnage of Mayari steels will surpass that of all the other chrome-nickel steels taken together.

CASTINGS OF NICKEL-CHROMIUM STEELS

Castings are made also of chrome-nickel steel and may be used in the annealed or heat-treated condition.

COMPOSITION AND PROPERTIES OF CHROME-NICKEL STEEL CASTINGS

Sample No.	Composition.							Tensile Properties.				Condition.
	C	Mn	Si	S	P	Ni	Cr	Tensile Strength.	Elastic Limit.	Contraction of Area.	Elongation in 2 Ins.	
	%	%	%	%	%	%	%	Pounds	Pounds	%	%	
1	0.30	0.41	....	0.04	0.03	3.64	1.49	91,500	45,550	24	16.5	Annealed
2	.33	.39	....	.04	.03	3.58	1.61	90,500	46,500	27	18.5	Annealed
3	.30	.20	0.35	....	....	2.50	.50	110,000	80,000	30	20	Heat-treated

BIBLIOGRAPHY ON NICKEL-CHROMIUM STEEL

Guillet, Léon. Aciers nickel chrome. Rev. mét., t. 3, 1906, pp. 462–484.

Lake, E. F. Nickel-chrome steel. Machinery (N. Y.), vol. 13, July, 1907, pp. 615–617.

Schaeffers, Joseph. Alloy-steel tubing for the automobile industry. Horseless Age., vol. 26, Dec. 7, 1910, p. 782.

Iron Age. Mayari steel; its properties and uses. Vol. 89, Jan. 4, 1912, pp. 69–72.

CHAPTER VII

SILICON STEELS

ALTHOUGH silicon is an ingredient of practically all steels its presence is often accidental or unavoidable, and if it is added to simple structural and tool steels the purpose is to promote soundness rather than to improve the properties of the finished steel. In tool steels silicon is always present, and in times past high-silicon steels have been advocated for tools, but they are not now so used in a commercial way.

MANUFACTURE OF SILICON STEEL

Silicon steels are generally made in the open-hearth furnace, preferably on an acid hearth, as the acid slag does not waste the silicon in the final additions as rapidly as does a basic slag that contains free oxide of iron, and therefore the final content of silicon desired may be more closely controlled.

Silicon in true silicon steels must be added to the charge only a short time before teeming, as any oxygen that reaches the metal will largely be taken up by the silicon which will be wasted by burning to silicic acid (SiO_2). When so added to a bath in proper condition as to temperature and amount of dissolved oxygen or oxides the silicon will overwhelm the gases in solution, and the steel as cast will be free from blowholes and with a maximum tendency to pipe.

Because of the large proportion of silicon in silicon steels and because of the short time allowable after the silicon has been added to the bath it should be added in the heated or molten state.

Silicon steels containing about 2 per cent of silicon or more roll very "dry," that is, they are liable to be cracked by the heavy reductions of the first passages through the blooming mill.

PROPERTIES OF SILICON STEELS

Silicon steel containing 0.20 per cent of carbon may be rolled if the silicon content is less than 7 per cent. With 0.90 per cent carbon it may be rolled if the silicon is less than 5 per cent. With a silicon content higher than 5 per cent the metal is useless.* In structural steels the effect of the silicon is to raise the elastic limit to a moderate degree. Silicon lowers the coefficient of expansion of steel somewhat as nickel does.

The treated test piece comprising sample 5 was heated to 954° C. (1,750° F.), quenched in water, and drawn at 427° C. (800° F.). The hardening temperature of samples 8 and 9 was probably about the same as that of sample 5.

USES OF SILICON STEELS

The dividing line between silicon-treated steels and silicon-alloy steels is not clearly defined, but the latter are used for several important purposes. In structural lines their employment is limited, as their properties can, generally speaking, be readily equaled or excelled by simple steels.

The chief structural use of silicon-alloy steel is in springs of the leaf type for automobiles and other vehicles. The silicon is considered to make the springs somewhat tougher, so that they are less liable to break in service than springs of simple steel. In the trade this steel is called silico-manganese steel, though its content of manganese is usually no more than is common in simple steels and not enough to properly cause the steel to be classified as a manganese-alloy steel.

In electricity, an important use for silicon-alloy steel is in the cores of static transformers. With the exception of manganese most of the elements employed in making alloy steels,

* Guillet, Léon, Aciers au silicium: Rev. mét., t. 1, 1904, p. 67.

COMPOSITION AND PROPERTIES OF STRUCTURAL SILICON ("SILICO-MANGANESE") STEELS

Sample No.	Description.	C	Si	Mn	S	P	Tensile strength.	Elastic limit.	Elon-gation.	Con-traction.	Ball hardness.
		Per Cent.	Per Cent.	Per Cent.	Per Cent.	Per Cent.	Pounds.	Pounds.	Per Cent.	Per Cent.	
1	Automobile springs	0.50	2.0	0.70	0.04	0.03	....	....	....	....	....
2	Springs treated	.47	1.83	.70	.01	.01	254,000	230,000	9	40	....
3	Carriage axles	.50	1.90	.70	.04	.04	....	....	....	....	....
4	Test piece, natural condition	.48	1.4	.45	....	....	113,760	71,100	17	....	....
5	Test piece, treated	.48	1.4	.45	....	....	177,750	149,310	14	....	....
6	Test, treated	.50	1.75	.65	....	....	....	198,700	8.5	21	418
7	Annealed	.36	1.27	.57	.03	.02	94,500	59,750	25	48	....
8	Drawn at 427° C.	.36	1.27	.57	.03	.02	182,200	160,850	12.5	34	....
9	Drawn at 427° C.	.31	2.39	.48	.03	.05	134,750	104,700	22	55	....

although not greatly decreasing the magnetic susceptibility of the iron that contains them, lower its hysteresis loss. Silicon is the element most used for that purpose because it is the cheapest, but aluminum, phosphorus, nickel ($3\frac{1}{2}$ per cent), and tungsten have a similar effect.

The silicon content in silicon transformer metal is usually between $3\frac{1}{2}$ and $4\frac{1}{2}$ per cent or, more exactly, 4 to $4\frac{1}{4}$ per cent. Some 25,000 tons was used in 1913 for this purpose. The steel is rolled into thin sheets which, for one large user, are 0.014 inch thick; the transformer cores are built up of these sheets, which are cut to shape separately by stamping. For low induction the permeability of this steel is nearly as great if not greater than that of any other variety of iron or iron alloy known, and its hysteresis loss is less than that of any other of nearly as low cost.

The results of an analysis of a transformer core made of silicon-alloy steel are as follows: C, 0.08 per cent; Si, 4.18 per cent; Mn, 0.11 per cent; S, 0.06 per cent; P, 0.01 per cent; Al, 0.01 per cent.

Silicon steels can not be case hardened, as the silicon retards the absorption of carbon; the silicon content must therefore be low, not over 0.04 per cent, in steel intended to be so treated.

BIBLIOGRAPHY ON SILICON STEEL

Hadfield, R. A. On alloys of iron and silicon. Jour. Iron and Steel Inst., 1889, pt. 2, pp. 222–255.

Heyn, E. Einfluss des Siliciums auf die Festigkeitseigenschaften des Flussstahls. Stahl und Eisen, Jahrg. 21, May 1, 1901, pp. 460–464.

Guillet, Léon. Aciers au silicium. Rev. mét. t. 1, 1904, p. 67.

Gontermann, W. Iron-silicon-carbon alloys. Jour. Iron and Steel Inst., vol. 83, 1911, pp. 421–475.

Ruder, W. E. Grain growth in silicon steel. Bull. Am. Inst. Min. Eng., December, 1913, pp. 2805–2822.

CHAPTER VIII

HIGH-SPEED TOOL STEELS

HIGH-SPEED tool steels, also called rapid steels, have in the past fifteen years worked a remarkable revolution in the machine-shop business of the whole world, affording largely increased outputs and commensurate lower costs. As a consequence they are now being used very generally and in some shops almost exclusively for machining iron and steel as well as some other metals by cutting operations by machine tools.

The revolutionary feature wherein tools made of these steels differ from and exceed in service the tools formerly used is their ability to maintain a sharp strong cutting edge while heated to a temperature far above that which would at once destroy the cutting ability of a simple steel tool. Because of this property a tool made of high-speed tool steel can be made to cut continuously at speeds three to five times as great as that practicable with other tools, and when, as the result of the friction of the chip on the tool, it may be red-hot at the point on top where the chip rubs hardest, and the chip itself may, by its friction on the tool and the internal work done on it by upsetting it, be heated to a blue heat of 296° C. (565° F.) or even hotter to perhaps 340° C. (644° F.).

This property of red-hardness or ability to retain hardness at a red heat may be imparted to steels of suitable composition, comprising chromium and tungsten, by a unique heat treatment to which they may be subjected. This treatment, described later, was introduced by F. W. Taylor and Maunsel White, as has been described by Taylor,* at the works of the Bethle-

* Taylor, F. W., On the art of cutting tools: Trans. Am. Soc. Mech. Eng., vol. 28, 1906, pp. 31–350.

hem Steel Co. in 1899, and tools so treated were shown at the Paris Exposition in 1900, where they naturally created a great sensation among those familiar with the machining of metals. White, when giving the writer his first knowledge of these tools in 1899 or early in 1900, said that a young man in the Bethlehem shop had lighted a cigarette with a newly cut chip, a statement that seemed almost unbelievable at the time.

In this country in 1913 about 7,000 tons of high-speed or rapid tool steels was made by some 15 makers, that output requiring about 8,000 tons of ingots.

MANUFACTURE OF HIGH-SPEED TOOL STEEL

High-speed tool steels are all made by the crucible or electric-furnace process. Except at one works, the crucibles or pots are made of graphite. The average life of the crucibles or pots varies in different works from six to nine melts. Some makers use clay-lined graphite pots in melting this steel to prevent or hinder the absorption of carbon from the pot. The clay lining is only one-eighth to three-sixteenths of an inch thick and is sometimes cut through on the second or third melt; in that event the molten steel may absorb too much carbon. Other makers use a graphite pot twice—first for melting other kinds of steel and then for rapid steel when the inner surface of the pot is somewhat slagged over, because of which the absorption of carbon is much less than when the pot was new.

The large producers use gas-fired melting furnaces for heating the pots, which are charged into the furnace at the top. Each melting hole contains six pots and each pot takes a charge of 90 or 100 pounds. The charge is melted and then "killed" in the usual way by being held 30 to 40 minutes. Such procedure, together with the presence of the large amount of alloy, regularly gives sound piping steel. If run continuously a furnaceful of pots will be melted about every four hours.

In packing a pot with a charge for rapid steel the tungsten must be placed on top of the charge—as with simple tungsten steel—to guard as far as possible against the tendency of the tungsten to settle because of its high specific gravity. That tendency seems to be less with the rapid steels than with the simple tungsten steels. Whether the chromium of the former influences or hinders the settling of the tungtsen is conjectural.

The smaller ingots, which are made from one pot of steel, vary from $3\frac{1}{2}$ to 5 inches square. The steel is sometimes teemed directly into the mold by hand-pouring, but in some works clay funnels are placed on top of the mold to direct the stream down the center of the mold to avoid cutting its wall, as might happen if the stream impinged on it. Funnel pouring is also advantageous when two pots are to be combined to make a larger ingot, as the steel can be poured into the funnel from opposite sides at the same time, a procedure that will mix the liquid steel and give a more uniform ingot than when one pot follows another, as in hand-pouring when no funnel is used.

Some of the larger producers of rapid steels use for casting a large bottom-pouring ladle into which the steel is poured from the pots of one or more furnaces, and from which the ingots are top-cast; that is, the molds are filled from the top. This method presents the advantages that (1) the product is more uniform; (2) individual pot charges, which might not be of the prescribed composition or might be otherwise unsatisfactory, are merged with the others without detriment to the whole; (3) large ingots are easily made; (4) one analysis serves for the whole number of pots; (5) one test serves for the whole ladleful of steel. It is a matter of experience that complaints from customers became much less frequent after the introduction of the ladle for casting this steel.

The strong tendency of rapid steel to pipe is checked considerably in most plants by the use on each ingot of a hot "dozzler," which is a clay ring preheated red hot, that is placed on the ingot top and filled with molten steel. This arrangement keeps the top of the ingot molten long enough so that the pipe is of diminished size and nearly or quite all contained within the part of the ingot surrounded by the dozzler. The proportion of the ingot to be rejected on account of the pipe is therefore much decreased. The molds are usually closed at the bottom end and are either made with parallel walls or tapered so that the ingot is larger at the top than at the bottom. The molds must be split when the walls are parallel, and are sometimes split when the ingots are tapered.

High-speed tool steel as cast has a coarse structure and dark color as compared with the structure and color of simple steels of the same carbon content. A corner is broken from the top of each ingot, to show the grain, and the ingots when hand-poured directly from the pots are classified by the eye as in the production of simple crucible steels. If the ingots are cast from the large ladle a test is taken for analysis which determines the disposition of the whole ladleful of steel.

As a rule the ingots show a strong columnar structure or arrangement of crystals whose axes are normal to the cooling surface. Some makers refer to the structure as a "lemon structure," the crystals of the metal being thought to resemble the cells forming the pulp of a lemon. If the casting temperature is lower than usual, this lemon structure may be absent, and in that case the interior of the ingot will have a much finer grain than the ingots cast at the usual higher temperature. The subsequent heating and working of the steel entirely destroys the crystalline structure of the ingot, and the worked steel, on a fresh fracture, shows a most beautiful porcelanic structure.

The ingots run from 3½ by 3½ inches to 16 by 16 inches, but most of them are from 5 by 5 inches to 9 by 9 inches. For hot-working they are heated in a furnace chamber having a temperature of about 1,180° C. (2,156° F.). At this high heat the steel may be worked satisfactorily under the hammer or press and may be quickly worked down to the dimension desired.

COMPOSITION OF HIGH-SPEED TOOL STEELS

The tendency of the makers is toward a somewhat uniform composition as regards the contents of the alloying elements, whose benefits have become fairly well known, and whose use as a consequence may be considered as established. Specifically, these alloying elements are tungsten and chromium. The addition of vanadium and cobalt in important proportions is considered by some makers to give distinct improvement to high-speed steel, and some vanadium is almost always present.

The following analyses are of steels recently made, most of which are considered to be good commercial steels:

RESULTS OF ANALYSES OF HIGH-SPEED STEELS MADE IN 1913 OR 1914.

Sample.[1]	C	Mn	Si	S	P	Cr	W	V	Co	Ni	Mo	Remarks.
	%	%	%	%	%	%	%	%	%	%	%	
A—	0.65	0.15	0.20	0.02	0.03	4.75	17.50	0.90	.....	.....	.....	
B—1	.66	.27	.14	.04	.05	4.51	17.48	.70	4.22	0.17	.....	
B—2	.74	.31	.13	.04	.02	4.20	15.63	.67	2.70	.....	.....	
B—3	.63	.14	.07	.04	.05	4.26	17.16	.45	3.80	.20	.....	
B—4	.69	.34	.14	.03	.04	5.28	16.35	.64	5.28	.....	.....	
C—1	.66	.22	.17	.03	.02	3.44	16.51	.73	.....	.....	.....	
C—2	.64	.21	.16	.03	.03	3.30	16.06	.66	4.02	.....	.....	
C—3	.67	.33	.25	.02	.02	3.85	16.06	.70	.....	.....	.....	
D—1	.75	.28	.36	.03	.....	4.10	19.00	.75	.....	.....	.....	Good
D—2	.68	.38	.40	.03	.....	4.65	17.85	.53	.....	.....	.....	Inferior
D—3	.69	.36	.38	.04	.....	4.67	17.90	.50	.....	.....	.....	Inferior
D—4	.57	.20	.26	.02	.03	4.82	15.38	.50	.....	.....	.....	Inferior
E—1	.61	.23	.35	.04	.....	4.10	17.20	1.00	.....	.....	.....	Good
E—2	.68	.45	.40	.04	.....	4.00	14.26	1.09	.....	.....	.....	Inferior
E—3	.70	.50	.39	.05	....	4.08	14.50	1.07	.....	.....	.....	Inferior
E—4	.60	.23	.12	.03	.02	3.90	17.27	.90	.....	.....	.....	Inferior
F	.64	2.29	.12	.02	.01	4.39	16.09	.59	.....	.28	.....	
G	.72	.37	.18	.03	.02	4.50	13.30	2.50	.....	.....	.....	
H—1	.77	.16	.21	.02	.02	4.05	18.64	1.35	.....	.....	.....	
H—2	.67	.16	.20	.02	.02	4.66	13.86	1.08	.....	.....	.....	
I	.64	.23	.29	.02	.02	4.57	19.10	.54	.....	.....	.....	
J—1	.64	.30	.26	.02	.01	2.93	18.71	1.22	.....	.....	.....	
J—2	.71	.14	.26	.03	.03	2.97	18.21	.97	.....	.....	.....	
K—1	.55	Tr.	.23	.02	.04	4 46	16.05	.80	4.72	.....	0.72	
K—2	.70	Tr.	.18	.01	.02	4.25	15.50	.88	4.72	.18	.67	
K—3	.74	.31	.13	.04	.02	4.20	15.63	.67	2.70	.....	.....	

[1] Samples A to I represented American steels, the numerals indicating different samples from the same maker; sample J represented an English steel; sample K represented a German steel.

Samples D—1 and E—1 gave excellent results in a competitive test, whereas samples D—2, D—3, E—2, and E—3, manufactured by the same makers, gave distinctly inferior results in the same shop.

The occurrence of nickel in four of the samples may have been accidental, having been due to nickel in some of the scrap steel used in the charge. Most makers now put in vanadium, and steel like that represented by sample G, which had the highest vanadium content of all the samples represented in the table, was the winner in a recent competitive test.

The average specific gravity of the steels represented in the table was about 8.8, the increase over the specific gravity of iron being due chiefly to the tungsten content.

There are so many factors beside the ultimate composition that affect the value of rapid tool steels that no conclusion can be drawn from the analysis alone. The melting, hot working, and heat treatment all must be done correctly or the final result will not conform to expectations.

CARBON IN HIGH-SPEED TOOL STEEL

The proportion of carbon aimed at in high-speed tool steels is about 0.65 per cent, which in a simple steel would not be enough to give the maximum hardness even if the steel were heated above the critical point and quenched in water, and still less so when the steel is cooled as slowly as these steels are in their treatment. This shows that the carbon acts in a different way from what it does in simple steels, as is discussed later.

TUNGSTEN IN HIGH-SPEED TOOL STEEL

Tungsten is well established as a most important if not indispensable ingredient of commercial tool steels, being almost or quite universally used in quantity therein. The best proportion of tungsten, all things considered, seems to lie between 16 and 20 per cent, the tungsten content in 95 per cent of all the American steels coming within these limits. Some published analyses of European high-speed tool steels show a higher content of tungsten than this, but American makers generally agree that any tungsten in excess of 20 per cent adds nothing to the usefulness of the steel, and they therefore make that proportion the upper limit of the amount added. One effect of the tungsten is that the best percentage of carbon in rapid steel is but about half that required in simple tool steels intended for the same kind of service.

CHROMIUM IN HIGH-SPEED TOOL STEEL

The effect of chromium in high-speed tool steel, as in other steels, is undoubtedly as a hardener, entering into the double carbide of tungsten and chromium which gives or causes the proper cutting edge. Although the proportion of this element present in these steels varies considerably, it is always large, perhaps never less than 2 per cent or more than 6 per cent in American steels, and in European steels the upper limits is at least 9 per cent.

MOLYBDENUM IN HIGH-SPEED TOOL STEEL

The use of molybdenum in high-speed tool steels is being generally discontinued. Some makers for years manufactured molybdenum tool steels, but as a rule they have either wholly discontinued its use or use a much smaller proportion than formerly, employing it as an auxiliary rather than a major constituent.

The effect of molybdenum is similar to that of tungsten, but is more intense in that 1 per cent molybdenum is currently considered to give about the same or greater hardening effect than 2 per cent of tungsten. It gives a fine cutting edge.

Various reasons are assigned for the discontinuance of the use of molybdenum in these steels. Taylor* found that molybdenum in rapid steels caused irregular performance; that steels of nearly the same composition and having had seemingly the same treatment gave large variations in the cutting speeds they would stand. One user specifies no molybdenum because it causes the tools to crack in quenching. A maker objected to molybdenum because molybdenum steel was apt to be seamy and to contain physical imperfections. A maker of ferro-alloys understands that molybdenum steel deteriorated upon repeated heating for dressing and treatment, seemingly because the molybdenum disappeared from the outer parts of the steel

* Taylor, F. W., On the art of cutting metals: Trans. Am. Chem. Soc., vol. 28, 1906, pp. 31–350.

where it was exposed to the heat and where any volatile constituent could escape. This phenomenon recalls a proposal of Moissan to use molybdenum as a means of freeing molten iron of oxygen, as he found that the oxide of molybdenum was much more volatile than the metal itself. It may be, therefore, that the molybdenum at the surface is oxidized and volatilized, but for this waste to extend to any considerable depth implies either that molybdenum moves through the heated steel to reach the oxygen of the air, or, as is more likely, that the oxygen of the air penetrates the steel, reaching the molybdenum and oxidizing it, the volatile oxide escaping from the steel.

The cost of molybdenum when its use was most general was much more than that of tungsten. Because of the falling off of the demand in recent years, however, the price of molybdenum has fallen to about that of tungsten, or to a fraction of what it was a few years ago.

It has been observed that in the Eggertz color determination a high molybdenum content (4 to 8 per cent) causes "missing" or invisible carbon when the content of latter element is high (0.4 to 1.2 per cent).

VANADIUM IN HIGH-SPEED TOOL STEEL

Vanadium is used for high-speed tool steel in varying amounts, most makers using at least 0.5 per cent, although some run the vanadium content up to $1\frac{1}{2}$ or $1\frac{3}{4}$ per cent, or even more, considering that such an addition increases in an important degree the value of the steel for tools.

The effect of vanadium is considered to resemble in some ways that of chromium in increasing the hardness or red-hardness of the cutting edge. One maker makes two kinds of high-speed tool steel, the essential difference between the two being that one contains about $1\frac{1}{2}$ per cent of vanadium and thereby commands a higher price, whereas the other has little of this element, both kinds being intended for substantially the same service.

High-speed steels containing vanadium are generally classed

as "superior" steels and many, though not all, makers and users consider them distinctly better than the "standard" steels containing no vanadium, both on account of their actual cutting qualities at high speeds and on account of the length of time a tool will cut before it needs regrinding. The true value of vanadium in rapid steels must probably be held as not yet fully determined. The analyses given in the table show that all the samples contained vanadium but in greatlv varying amounts.

COBALT IN HIGH-SPEED TOOL STEEL

Cobalt now threatens to change tool-steel manufacture because of the properties it imparts. The recent great decline in price following the increase of the supply from the silver ores of the Cobalt district in Ontario naturally led to its trial as a steel-alloying element, and some most excellent high-speed steels containing, in addition to the usual ingredients, about 4 per cent of cobalt, have been obtained. This result was hardly to have been expected in view of the experience with nickel, which cobalt much resembles, as nickel has been condemned by nearly every manufacturer as not being a desirable ingredient of high-speed tool steels, because of the effect it has of making the edge soft or "leady." The cobalt steel, however, has shown, in some products at least, increased ability to hold its edge in work.

One user of cobalt steel found it better suited for turning manganese steel than any other steel he tried, his success being so marked as to make it practically a commercial operation. Manganese steel, as noted elsewhere, is so hard as to be considered practically unmachinable, the usual practice having been to finish it by grinding when necessary. Another user found that an imported cobalt high-speed steel proved excellent for cutting a hard nickel-chromium steel in a lathe, whereas the same steel in a cold saw was not satisfactory. The direct reason seemed to be that this steel to do its best work must be run nearly if not quite red hot; at least that was the condition in which it was used in the lathe, while in the cold saw its temperature was not raised to an important degree, and when cold its properties were different than when it was hot. This behavior is more or less common to high-speed steel, as is mentioned later.

The valuable effect of cobalt is claimed to be that it increases the red-hardness of high-speed tool steel, enabling the steel to cut at a higher speed.

COPPER IN HIGH-SPEED TOOL STEEL

Copper has been considered to be highly injurious in high-speed tool steel, even as little as 0.05 per cent being inadmissible; and it is thought to be particularly harmful if much sulphur is present in the steel; also the higher the carbon content the more harmful is the copper.

SULPHUR AND PHOSPHORUS IN HIGH-SPEED TOOL STEEL

Sulphur and phosphorus, which are so deleterious in simple tool steels, are considered to be somewhat less so in high-speed steels, in which the effect is either modified or else masked by the large quantities of other ingredients. Some commercial brands of high-speed steels have as much as 0.05 per cent of each of these impurities, to which no inferior quality is attributable.

STELLITE

Stellite, though a competitor of high-speed steels, is not within the scope of our subject, but a recent analysis is given of a sample for such interest as it may have in relation to cutting steels.

ANALYSIS OF STELLITE

Constituent.	Per Cent.
Co	59.50
Cr	10.77
Mo	22.50
C	.87
Si	.77
Mn	2.04
S	.084
P	.040
Fe	3.11
W	0
Ni	0
	99.684

HEAT TREATMENT OF HIGH-SPEED TOOLS

The heat treatment given to high-speed steels for the commoner uses as lathe and planer tools has generally been simplified to heating to incipient fusion and quenching in oil. Cooling by an air blast and double treatment, which were formerly recommended, are now not common, except that a second (drawing) heating is given to milling cutters and similar tools, the temperature imparted to the tool depending on the material to be cut.

The treatment is usually done by the blacksmith who heats the tool in his forge fire and then immerses it in a tank containing enough oil so that its temperature does not rise materially. Ten gallons of oil is a common quantity to use when the size and number of the tools is moderate, as in most shops. The fire is a deep compact coal fire, the coal in the center where the tool is heated being pretty thoroughly coked, that is, most of its volatile matter distilled out. This manner of heating has the advantage that free oxygen does not get at the tool to oxidize it, but its environment is nonoxidizing or even reducing, owing to the presence of an excess of burning carbon surrounding the tool. Any flame is more or less oxidizing, at least unless heavily charged with smoke or free carbon, and a piece of steel heated directly by a flame as in the ordinary heating chamber of a furnace is likely to be somewhat oxidized on its surface, the depth to which the oxygen penetrates varying according to the conditions, particularly the temperature, the access of air, and the length of time. Heating in a muffle will also result in oxiizing the steel unless extraordinary precautions are taken to keep out oxygen or to consume all that enters. The temperature of quenching, usually about 1,260° C. (2,300° F.), is determined by the fusion of the scale and its visible collection into drops or beads on the surface of the tool.

Quenching is done by quickly plunging the heated tool into the oil as soon as it has reached the desired temperature and moving it about in the oil until cold. Cooling in oil is thought by some to give a better tool than cooling in the air blast, one reason seemingly being the protection of the steel from free oxygen while it is hot enough to be oxidized thereby. The oxygen of the air blast forms a scale of oxide on the hot steel and the oxygen probably penetrates the metal below the scale to some extent, injuring the quality as deep as it goes. A tool on its second grinding when the oxidized metal is removed may then give better service than on the first, unless the first grinding has for that reason been heavy enough to remove the oxidized metal.

In some shops, however, the original treatment recommended by Taylor and White* is given, the cutting edge of the tool being heated to incipient fusion and then immersed in a bath of melted lead at about 565° C. (1,050° F.). The heating is done in a small furnace over a deep coke fire, blown by an air blast, so that the environment of the tool while being heated is substantially nonoxidizing. Flames of carbonic oxide play out of the openings through which the tools are inserted, indicating little if any free oxygen within. In these shops, however, milling cutters and other tools that are machined to a particular form are treated by heating them to a slightly lower temperature, in order not to damage the cutting edges, and then plunging them into cold oil.

When cooled to the temperature of the lead it is taken out and placed in an air blast to complete the cooling. Some tools desired to be especially tough so as not to break in service are given a second heating to 565° C. and then cooled in the open air or air blast if saving time is important.

Rapid steel when well annealed will bend considerably without breaking, even in as large a section as $2\frac{1}{2}$ by $1\frac{1}{4}$ inches, the bending being edgewise, as in a tool at work.

Gledhill† found that one of these steels after having been annealed 12 to 18 hours at 760° C. (1,400° F.) had a tensility of 129,200 pounds per square inch, an elastic limit of 89,600

* Taylor, F. W., On the art of cutting metals: Trans. Am. Soc. Mech. Eng., vol. 28, 1906, p. 228.

† Gledhill, J. M., The development and use of high-speed tool steel: Jour. Iron and Steel Inst., pt. 2, 1904, pp. 127–181.

pounds per square inch, an elongation of 18 per cent in 2 inches, and a contraction of area of 35 per cent. The ductility is rather high and would enable a tool to be bent considerably without breaking. Such annealed steel may be rather easily machined for making milling cutters and other shapes that require machining.

Carpenter* found that the higher the temperature from which rapid steel is cooled the more it resisted etching for metallographic work. He also found that no tempering change occurred when it was reheated at a temperature of less than 550° C. (1,022° F). to a visible red in the dark, indicating a stability that is doubtless the cause of its property of red-hardness.

Whether a rapid steel is made harder by the heat treatment given it depends somewhat on the condition of the bar before treatment. If it has previously been annealed, the treatment hardens it, whereas heat treatment may not harden a piece in the natural state. Taylor† found that some tools having useful red-hardness could be filed rather readily. Edwards ‡ on the other hand found treated high-speed steels to be exceedingly hard—as hard as any steel could be made by quenching. Gledhill § found that high-speed steel was good for turning chilled rolls which are extremely hard and require the hardest kind of tool to cut them.

Trials on window glass of a number of different rapid steels showed that the cutting edge of some but not of all would scratch it. The same was true of the untreated ends of the same tools, as some would and some would not scratch the window pane.

The hardness of the steel when cold is not the determining factor of usefulness in any case. It is the hardness when heated under conditions of work.

* Carpenter, H. C. H., The types of structure and the critical ranges on heating and cooling of high-speed tool steels under varying thermal treatment: Jour. Iron and Steel Inst., pt. 1, 1905, pp. 433-473.

† Taylor, F. W., On the art of cutting metals: Trans. Am. Soc. Mech. Eng., vol. 28, 1906, pp. 31-350.

‡ Edwards, C. A., Function of chromium and tungsten in high-speed tool steel: Jour. Iron and Steel Inst., pt. 2, 1908, pp. 104-132.

§ Gledhill, J. M., loc. cit.

The cutting edge of a rapid-steel tool at work is probably never as hot as the metal just back of it, where the heating caused by the friction of the chip, as it is deflected and rubs hard on the tool, is most intense. The edge itself is kept relatively cool by the cold metal flowing upon it.

THEORY OF HIGH-SPEED STEELS

Carpenter* found the heating and cooling curves of a rapid steel to be radically different from each other, and also that the cooling curve when the steel was cooled from 930° C. (1,706° F.) was greatly different from that when the steel was cooled from 1,250° C. (2,282° F.). When the steel was cooled from 930° C. the curve had an abrupt jog, which showed a great retardation in rate of cooling occurring between 700° C. and 750° C. (1,292° F. to 1,382° F.). The jog did not occur when the steel was cooled from 1,250° C., 320° higher, the line representing variations in rate of cooling being nearly straight. The rate of cooling to get these curves was slow or at least not accelerated, and one can not say what the curves would be like if the rate of cooling were hastened, as in quenching, but the curves obtained seem to throw much light on the question. The property of red-hardness seems to be connected with the elimination of the great retardation mentioned.

The following explanation, based on the work of Carpenter* and Edwards†, of the properties of high-speed steels, seems to be helpful or even satisfactory:

Their researches on the heating and cooling of these steels have shown that such steels have an extraordinary stability of composition after they have been heated to 1,200° C. (2,192° F.) or more, and that a second heating of 550° C. (1,022° F.) has no softening or drawing effect. It seems fairly evident that red-hardness depends on or is the natural result of these facts.

At a temperature higher than 1,200° C. (2,192° F.) a double carbide of chromium and tungsten is formed, which persists largely even when the steel is cooled slowly as in the open air,

* Carpenter, H. C. H., loc. cit. † Edwards, C. A., loc. cit.

and more so when cooling is accelerated. This double carbide imparts to the steel a high degree of hardness and is stable at all temperatures up to 550° C. (1,022° F.) or somewhat higher. At 550° C. the steel has a low red color visible in the dark.

If the above theory be true then at a temperature of 1,200° C. (2,192° F.) the chromium and tungtsen must have a stronger affinity for carbon than iron has, whereas at lower temperatures, say from around 930° C. down to the critical point, the affinity of carbon for iron is slightly stronger than that of either chromium or tungsten or both, and the carbon then exists wholly or in part as carbide of iron, or a complex carbide of iron with one or both of the other elements.

Carbide of iron, or hardening carbon which causes the hard condition of iron in simple steel that has been quenched from a temperature higher than the critical point, is unstable at even slight elevations of temperature above atmospheric temperature, its unstableness increasing with the degree of heat, though not being proportional thereto. Boynton* has shown that between 400° C. (752° F.) and 500° C. (952° F.) the amount of change and consequent softening is much greater than at other temperatures, either lower or higher.

The proportion of carbon in rapid steel should perhaps be only as much as will combine with the chromium and tungsten at 1,200° C. (2,192° F.) and leave none to exist as unstable hardening carbon of hardened simple steel.

TESTING AND USING HIGH-SPEED STEEL

A reliable and inexpensive method of quickly testing high-speed steels to show their value is much needed, as Taylor† has explained. Herbert‡ and Edwards§ have used and recom-

* Boynton, H. C., Hardness of the constituents of iron and steel: Jour. Iron and Steel Inst., 1906, p. 287.

† Taylor, F. W., On the art of cutting metals: Trans. Am. Mech. Eng., vol. 28, 1906, pp. 31–350.

‡ Herbert, E. G., The testing of files and tool steels: Trans. Manchester Assn. Eng., 1908–1909, pp. 302–317.

§ Edwards, E. T., Composition of high-speed tool steel: Iron Age, vol. 89, April, 1912, pp. 957–960.

mended machines and methods that lessen the time and trouble of testing, but no test seems to take the place of a trial at actual work, because the performance of a tool in one line of work with certain conditions may not be foretold positively by its performance in another with different conditions. Among the reasons are that (1) sometimes greater durability is obtained by changing, that is, increasing or lessening, the speed of the cut, thus changing also the temperature of the tool, or (2) a given tool when used at its best speed may be excellent for cutting a certain material, yet prove inferior to another tool for cutting a different material. Thus if selected as the best by trial for cutting a 0.20 per cent carbon steel, it may be surpassed by others in cutting a 0.70 per cent carbon steel.

Physical tests of rapid steels at different temperatures up to 800° C. (1,472° F.) are needed to show the effect of heat on the physical properties of those steels. New uses would probably be suggested by the results of such a series of tests.

A rapid-steel tool does not finish the piece being cut as nicely as does a simple steel tool, as the rapid steel does not keep a fine edge with a light cut and slow speed of, say, 20 feet per minute. The durability of such a tool taking a light cut is much greater at a higher cutting speed, at which the tool is hotter, showing that the strength or the toughness of the steel or both are augmented by the higher temperature. Unhardened simple steels with c.6 to 0.7 per cent carbon get stronger but less ductile with a rise of temperature up to about 300° C. (572° F.). If, as the temperature rises, high-speed steels get stronger without loss of ductility but perhaps with an increase, within limits of course, a physical reason for their great durability is provided.

In 1910 Herbert* announced the discovery that any rapid-steel tool and some simple steel tools may have two rather widely separated cutting speeds at which the tool is more durable than at speeds above, below, or between. Thus out of many cases described, one tool cooled in an air jet had nearly equal

* Herbert, E. G., The cutting properties of tool steel: Jour. Iron and Steel Inst., 1910, pt. 1, p. 216.

maximum durability at two speeds—50 and 90 feet per minute, whereas at 65 feet the durability was less than one-half of that at either of the other speeds. This discovery no doubt accounts for some of the anomalies encountered in tool steels as well as other steels the properties or performances of which are not what would be expected from their composition and other attributes. Thus a tool may be condemned when an increase of its cutting speed would cause it to give satisfactory service and durability.

Rapid steel will do its best cutting when hot. A desirable practice, followed in some shops, is to heat a tool to near redness before putting it to work.

MACHINE-TOOL DESIGN

When Taylor and White first introduced rapid steels it was thought that the higher cutting speeds afforded constituted the sole benefit to be derived from them, and as the higher speeds, although consuming more power about in proportion to the increase in speed, did not increase materially the stresses on the machine tools, it was thought that the latter merely needed to be speeded up in order to get the full benefit from the new steels. But it was soon found that the rapid steels in addition to cutting at higher speeds were capable of taking much heavier cuts, which proportionately increased the stresses on the tools. To take full advantage of the heavier as well as the more rapid cuts, machine tools were generally redesigned to provide the greater strength required, and were supplied with proportionately more power. The resulting economies all over the world have been enormous.

The advantages from the use of rapid steels as compared with the use of simple tool steels are in the lessened costs of the ordinary operations of finishing iron and steel because of:

1. More rapid cutting speed.
2. Heavier chips cut, hence larger cuts and feeds.
3. Saving of power per unit of metal cut off.
4. Lower cost of plant for a given output.

5. Lower general and overhead charges connected with manufacturing iron and steel products.

PATENTS ON HIGH-SPEED STEELS

Since the original Taylor and White patents for treatment of high-speed steels were issued in 1901, others have been granted for almost every possible combination of elements which were in any way thought to be useful or valuable constituents of tool steel. Chromium and tungsten were generally included, though not always. Nevertheless most makers now aim at substantially the same foundation composition, with varying amounts of vanadium and sometimes with cobalt.

MISCELLANEOUS USES OF HIGH-SPEED STEELS

An important use for high-speed steel is in the exhaust valves for automobile engines, where it has given excellent results. These valves operate sometimes at a red heat and seemingly the property of red-hardness that the steel possesses enables it to give good service in these valves.

High-speed steel is being used also in the manufacture of extruded brass to form the die through which the extruded metal is forced. The temperature of the brass is high, near its fusion point, and seemingly the red-hardness of the steel enables the steel to perform this service satisfactorily.

A good file or a good cold chisel may be made of rapid steel, but they are not good enough to justify their cost as compared with those made of simple steels.

BIBLIOGRAPHY ON HIGH-SPEED TOOL STEEL

ENGINEERING. Rapid tool steels. Vol. 76, Aug. 21, 1903, pp. 255–256.

GLEDHILL, J. M. The development and use of high-speed tool steel. Jour. Iron and Steel Inst., 1904, pt. 2, pp. 127–181.

MARKHAM, E. A. The use of high-speed steels. Am. Machinist, vol. 27, pt. 1, Apr. 7, 1904, pp. 443–444.

NICOLSON, J. T. Experiments with a lathe-tool dynamometer. Proc. Inst. Mech. Eng., 1904, pts. 3–4, pp. 883–935.

PENDLEBURY, C. Notes on tests of rapid-cutting steel tools. Engineer (London), vol. 97, Apr. 1, 1904, pp. 331–332.

CARPENTER, H. C. H. The types of structure and the critical ranges on heating and cooling of high-speed tool steels under varying thermal treatment. Jour. Iron and Steel Inst., pt. 1, 1905, pp. 433–473.

NEWBOLD, S. The Taylor-Newbold saw. Foundry, vol. 27, November, 1905, pp. 118–121.

CARPENTER, H. C. H. Tempering and cutting tests of high-speed steels. Jour. Iron and Steel Inst., pt. 3, 1906, pp. 377–396.

CLARAGE, E. T. The manufacture of tool steel. Am. Machinist, vol. 29, pt. 2, Nov. 1, 1906, pp. 573–576. An address delivered before the Northwest Railway Club.

TAYLOR, F. W. On the art of cutting metals. Trans. Am. Soc. Mech. Eng., vol. 28, 1906, pp. 31–350.

AUCHY, GEORGE. The theory of high-speed tool steel. Iron Age, vol. 80, Dec. 26, 1907, pp. 1818–1822.

CARPENTER, H. C. H. An analysis of the evolution of modern tool steel. Engineering, vol. 83, 1907, pp. 569, 633.

EDWARDS, C. A. Function of chromium and tungsten in high-speed tool steel. Jour. Iron and Steel Inst., pt. 2, 1908, pp. 104–132.

VALENTINE, A. L. Making and using of high-speed steel tools. Am. Machinist, vol. 31, pt. 2, July 2, 1908, pp. 6–9.

CARPENTER, H. C. H. Possible methods of improving modern high-speed turning tools. Trans. Manchester Assn. Eng., 1908–1909, pp. 81–120.

HERBERT, E. G. The testing of files and tool steel. Trans. Manchester Assn. Eng., 1908–1909, pp. 317–402.

ENGINEER (London). High-speed tool steels. Vol. 107, 1909, pp. 289, 347. Describes results of tests at Sheffield works.

BECKER, O. M. High-speed steel. 1910. 360 pp.

BERG, C. P. Heat treatment of high-speed tools. Jour. West. Soc. Eng., vol. 15, 1910, pp. 738–764.

BRACKENBURY, H. I. High-speed tools and machines to fit them. Proc. Inst. Mech. Eng., 1910, pts. 3 and 4, pp. 929–955, 986–1026.

SMITH, W. G. High-speed steel and its heat treatment. Mech. Eng. (Manchester), vol. 25, May 6, 1910, pp. 537–540.

BREARLEY, HARRY. The heat treatment of tool steel. London, 1911.

GLEDHILL, J. M. The practical use of high-speed tool steel. Mech. Eng. (Manchester) vol. 27, June 23, 1911, pp. 762–766.

SULLIVAN, W. B. Tool steel. Proc New York Railroad Club, vol. 22, 1911, pp. 2595–2633.

EDWARDS, E. T. Composition of high-speed tool steel. Iron Age, vol. 89, April 18, 1912, pp. 957–960.

HAMMOND, E. K. The manufacture of tool steel. Iron Age, vol. 90, Oct. 3, 1912, pp. 766–771.

HERBERT, E. G. The influence of heat on hardened tool steels, with special reference to the heat generated in cutting operations. Jour. Iron and Steel Inst., 1912, pt. 1, pp. 358–377.

KENNEY, L. H. Tool steel for the United States Navy. Trans. Soc. Naval Arch. and Marine Eng., vol. 20, 1912, pp. 345–574.

MACHINERY. Steels for taps, drills, and milling cutters. Vol. 19, November, 1912, pp. 181–182.

SULLIVAN, W. B. A study of the proper hardening and classification of tool steels. Proc. Railway Club of Pittsburgh, vol, 11, April 26, 1912, pp. 155–185. Gives results of drilling tests.

BIGGER, C. M. Tool steel from a salesman's point of view. Iron Age, vol. 91, March 20, 1913, pp. 706–708.

RIPPER, WILLIAM, and BURLEY, G. W. Cutting power of lathe turning tools. Proc. Inst. Mech. Eng., 1913, pts. 3 and 4, pp. 1067–1210.

SCHLESINGIER, G. Die Fortschritte deutscher Stahlwerke bei der Herstellung hochlegierter Schnellarbeitsstähle. Stahl und Eisen, Jahrg. 33, June 5, 1913, pp. 929–939.

CHAPTER IX

CHROMIUM-VANADIUM STEELS

CHROMIUM-VANADIUM steels, usually called in the trade chrome-vanadium steels, are the latest development in structural alloy steels that have gained an extensive market. In 1913 about 90,000 tons of ingots is thought to have been made, of which about 75,000 tons was sold in rolled and forged products. These steels are almost all made in the open-hearth furnace, the chromium and vanadium alloys being added shortly before casting.

The hot working of chrome-vanadium steels presents no especial difficulties. The total amount of alloying elements is not large in the commercial grades, and the steel acts in the press and rolls much like simple steels with somewhat higher carbon contents.

Chrome-vanadium steels are in their physical properties much like chrome-nickel steels, but they have a greater contraction of area for a given elastic limit than the latter.

This higher contraction of area in the pulling test seems in some way to be associated with machinability, as chrome-vanadium steel with an elastic limit of 150,000 pounds per square inch may be machined rapidly, whereas a chrome-nickel steel having such an elastic limit would quickly dull the cutting tool if cut at the same speed.

The greater part of the chrome-vanadium steels made goes into automobiles. They are preferred by some because of their greater freedom from surface imperfections, notably seams, which steels, containing nickel are prone to have if the ingots are at all unsound. Vanadium is a deoxidizer, whereas nickel is not, so that vanadium, when present, favors quality, and the smaller proportion required enables it to compete with nickel even though its cost is five or six times as great.

Chrome-vanadium steels are nearly always used in the heat-treated condition, but there are exceptions even in automobiles, as some frames, forgings, and shafts are made of the steel in its natural state. When heat-treated these steels are both hardened and drawn at slightly higher temperatures than are used with nickel-chromium steels to get similar properties. These temperatures are given in the table of heat-treated chrome-vanadium steels.

Some chrome-vanadium steel is said to be used in armor plate of medium thickness (4 inches), which is not face-hardened but has high properties imparted by heat treatment. Some such steel is used in high-duty forgings and structural parts of machines.

COMPOSITION AND PROPERTIES OF CHROME-VANADIUM STEELS IN NATURAL STATE

Sample No.	Composition.							Tensile Properties.				
	C	Mn	Si	S	P	V	Cr	Tensile Strength	Elastic Limit.	Contraction of Area.	Elongation in 2 Inches.	Ball Hardness.
	%	%	%	%	%	%	%	Pounds	Pounds	%	%	
1	0.57	0.84	0.27	0.03	0.01	0.31	1.36	98,000	75,750	68.5	28.1	175
2	.46	.48	.20	.02	.01	.14	1.17	82,250	52,500	71.0	34.0	160
3	.18	.32	.18	.02	.01	.20	.74	60,500	42,900	75.0	43.0	133
4	.30	.65	.10	.04	.04	.18	.90	....	45,000	69.0	35.0	[1] 155

[1] Annealed.

EXAMPLE OF SATISFACTORY USE OF CHROME-VANADIUM STEEL

A hydroelectric plant had shafts $6\frac{1}{2}$ inches in diameter, which transmitted 3,000 kw. each at 480 revolutions per minute, and all broke in service. The shafts were made of untreated nickel steel having an elastic limit of about 40,000 pounds per square inch. To make stronger shafts by increasing their size not being practicable, other shafts were made under the specification that the elastic limit of the steel should be at least 105,000

COMPOSITION AND PROPERTIES OF CHROME-VANADIUM STEELS IN HEAT-TREATED STATE

Sample No.	Composition.							Tensile Properties.					Treatment.[1]
	C	Mn	Si	S	P	Cr	V	Tensile Strength.	Elastic Limit.	Contraction of Area.	Elongation in 2 Inches.	Ball Hardness.	
	%	%	%	%	%	%	%	Pounds	Pounds	%	%		
1	0.30	0.65	0.10	0.04	0.04	0.90	0.18	......	101,000	64	20	255	899° W; 704° A
2	.30	.65	.10	.04	.04	.90	.18	......	180,400	43	10	430	899° W; 454° A
3	.30	.65	.10	.04	.04	.90	.18	......	200,000	52	10	429	899° W; 315° A
4	.28	.45	.26	.02	.01	1.00	.18	96,500	79,000	75	34	187	899° O; 676° A
5	.40	.75	.26	.01	.01	1.00	.17	148,000	120,000	53	20	270	926° O; 676° A
6	.40	.75	.26	.01	.01	1.00	.17	221,000	200,000	48	11	435	926° O; 426° A
7	.57	.37	.20	.02	.01	.69	.22	188,200	177,500	57	14	330	——; 426° A
8	1.06	.36	.22	.02	.02	.95	.11	135,550	126,750	49	21	248	——; 648° A.
9	.41	.49	.12	.03	.03	1.09	.11	86,900	77,250	70	33	152	——; 754° A
10	.25	.50	.10	.03	.02	.95	.75	131,700	113,100	56	18	....	....

[1] The first temperature given for each sample is that at which the steel was quenched, and the second the drawing temperature; W, O, and A represent water, oil, and air, the three cooling media used. Samples 8, 9, and 10 were hardened before being drawn at the temperatures given.

pounds per square inch, its contraction of area 40 per cent, and its ball hardness uniform within 5 per cent. Shafts to meet such qualifications were made of chromium-vanadium steel containing 0.33 per cent C, 0.54 per cent Mn, 0.022 per cent P, 0.030 per cent S, 0.89 per cent Cr, and 0.24 per cent V. The ingot, which was 30 by 25 inches in section, was rolled to an 18 by 18 inch bloom or billet, and the shafts were forged therefrom. The shafts were heat-treated, and a test from one of them, about the average of all those made, pulled at Watertown Arsenal on a 2-inch by 0.505 diameter section, gave results as follows:

RESULTS OF TESTS OF HEAT-TREATED CHROME-VANADIUM STEEL SHAFT

Elastic Limit.	Tensile Strength.	Elongation.	Contraction.	Ball Hardness.
Pounds	Pounds	Per Cent	Per Cent.	
105,260	127,310	15	46.2	278
......	......	......	......	283
......	......	......	......	278

These shafts met the specifications and proved satisfactory in service.

BIBLIOGRAPHY ON CHROMIUM-VANADIUM STEEL

SANKEY, H. R., and SMITH, J. K. Heat-treatment experiments with chrome-vanadium steel. Proc. Inst. Mech. Eng., 1904, pts. 3 and 4, pp. 1235–1317.

LAKE, E. F. Some properties of vanadium steel. Am. Machinist, vol. 30, pt. 1, May 2, 1907, pp. 632–634. Gives results of tests of chrome-vanadium steel.

GIBBS, W. E. Vanadium steel. Cassier's Mag., vol. 38, June, 1910, pp. 174–181.

SMITH, J. K., and TURNER, W. L. Vanadium and metallurgy. Iron and Coal Trades Rev., vol. 81, 1910, pp. 389, 463, 585, 663.

IRON TRADE REVIEW. Chrome-vanadium tire tests. Vol. 51, Dec. 26, 1912, pp. 1217–1218.

MITCHELL, A. F. Heat treatment of chrome-vanadium steel. Proc. Railway Club of Pittsburgh, vol. 12, Dec. 19, 1912, pp. 38–66.

IRON AGE. Chrome-vanadium rolled-steel wheels. Vol. 92, Nov. 13, 1913, p. 1106.

RECOVERING ALLOYED ELEMENTS

As the alloying elements of alloy steels are all more valuable than iron, some of them very much so, makers of alloy steels wish, of course, to save such elements in their alloy-steel scrap, if practicable without too great cost. The makers can save tungsten, cooper, nickel, and cobalt, each of which has an affinity for oxygen and a heat of combustion with oxygen less than the similar properties of iron; however, the makers can not at present save the manganese, silicon, vanadium, titanium, calcium, magnesium, and chromium, each of which has an affinity for oxygen and a heat of combustion greater than the similar properties of iron. Plans have been proposed for saving some of them. Saving a part of these oxidizable elements in reworking the scrap is in some instances of no benefit, because in replacing the wasted part by the addition of a fresh quantity of the ferro-alloy saturated with carbon there will be too much carbon in the finished steel. This is particularly true of manganese in manganese-steel scrap. The carbon in the alloy-steel scrap is protected from oxidation by the alloying element while the latter is being oxidized. With low-carbon alloy available the case is different, as the content of carbon in the steel may then be controlled.

CERIUM PYROPHORIC ALLOY

The pyrophoric metal used on cigar lighters and for igniters in miner's lamps might be considered as an alloy steel, as it consists substantially of 30 per cent of iron with 70 per cent of cerium. It was patented by Welsbach, whose name is identified with the Welsbach light. The striker is of the grade of hardened file steel with about 1.50 per cent carbon. The detached particles of the cerium-iron alloy take fire in the air, ignition being quickened no doubt by the heat generated in the impact of the striker.

CONCLUSION

Further advance in the development of new alloy steels as well as many new applications of those alloy steels already established, are to be expected. Trials are continually being made of new alloys of promise, some of which will doubtless win place in the list of useful alloy steels. Hadfield's iron alloy containing 5 per cent manganese and 15 per cent nickel, although not at present of use, may become so in the future, as its properties are rather remarkable.

As some of the alloys in steel, as well as any heat treatment it may have received, affect the carbon contained so that its effect in the color determination is changed, the regular practice in some steel-works laboratories is to make all carbon determinations by direct combustion of the whole sample with oxygen. This procedure avoids the uncertainties and errors of the color determination in analyzing heat-treated alloy steels.

INDEX

(Over)

CIVIL ENGINEERING—*Continued*

5e Highways; Municipal Engineering; Sanitary Engineering; Water Supply. **Forestry. Horticulture, Botany and Landscape Gardening.**

6—Design. Decoration. Drawing: General; Descriptive Geometry; Kinematics; Mechanical.

ELECTRICAL ENGINEERING—PHYSICS

7—General and Unclassified; Batteries; Central Station Practice; Distribution and Transmission; Dynamo-Electro Machinery; Electro-Chemistry and Metallurgy; Measuring Instruments and Miscellaneous Apparatus.

8—Astronomy. Meteorology. Explosives. Marine and Naval Engineering. Military. Miscellaneous Books.

MATHEMATICS

9—General; Algebra; Analytic and Plane Geometry; Calculus; Trigonometry; Vector Analysis.

MECHANICAL ENGINEERING

10a General and Unclassified; Foundry Practice; Shop Practice.

10b Gas Power and Internal Combustion Engines; Heating and Ventilation; Refrigeration.

10c Machine Design and Mechanism; Power Transmission; Steam Power and Power Plants; Thermodynamics and Heat Power.

11—Mechanics.

12—Medicine. Pharmacy. Medical and Pharmaceutical Chemistry. Sanitary Science and Engineering. Bacteriology and Biology.

MINING ENGINEERING

13—General; Assaying; Excavation, Earthwork, Tunneling, Etc.; Explosives; Geology; Metallurgy; Mineralogy; Prospecting; Ventilation.

Printed in Dunstable, United Kingdom

91221357R00067